Mickey
34
Aluminum

INDIANAPOLIS

RACING MEMORIES

1961-1969

Dave Friedman

Motorbooks International
Publishers & Wholesalers

First published in 1997 by Motorbooks International Publishers & Wholesalers, 729 Prospect Avenue, PO Box 1, Osceola, WI 54020-0001 USA

Library of Congress Cataloging-in-Publication Data Available

ISBN 0-7603-0142-5

On the front cover: Bobby Unser whirls around the track at the 1968 Indianapolis 500.—*Bob Tronolone*

On the frontispiece: Jim Clark led all but 10 laps (Foyt led those) of the 1965 race. Clark won the race with a record average speed of 150.686 miles per hour and finished almost two minutes ahead of second place Parnelli Jones.—*Dave Friedman*

On the title page: Gurney pits for fuel and tires during the 1962 Indianapolis 500. At this point the car was running in the top ten. —*Dave Friedman*

On the back cover: (top)Some of the many people who gather in the infield annually to watch the race and celebrate. (Middle) During the 1963 race, Art Malone (75) struggles with clutch trouble as he is lapped by Jim Hurtubise (56), A.J. Foyt (2), Jim Mc Elreath (8) Bobby Marshman (5), Rodger Ward (1), Don Branson (4), and Paul Goldsmith (99) early in the race. Malone made pit stops on laps two, three, and four to try and rectify the clutch problems. He finally retired on lap 18.—*Dave Friedman*

Printed in Hong Kong

Contents

ACKNOWLEDGMENTS

THIS book would not have been possible without the help of several wonderful people. Tim Parker and Michael Dapper of Motorbooks International backed this project from the beginning and, without their help and beliefs, this book would never have happened. Thanks also to my editor, Anne McKenna.

Five of the key players in the winds of change in Indianapolis during the 1960s took time to share their memories with me. Jack Brabham, John Cooper, Dan Gurney, Parnelli Jones, and Rodger Ward have a wonderful capacity to remember what happened 30 years ago and can tell it in a reasonably humorous manner. Their time and insight is much appreciated.

Susan Claudius corrected my grammar and spelling errors. Her findings made me realize that I shouldn't have slept through my English classes at Beverly Hills High School. She is also there when I need a little love and humor in my life.

All of the photographs in this book are from my collection and all of the facts are taken from the official races records, Floyd Clymer's *Indianapolis Yearbooks*, and *Autosport*.

A special note of thanks goes to former sports car champion Ed Leslie who allowed me to hide away in his beautiful Carmel Valley home and get this project done without any outside interference.

Again, thanks to everyone who has been part of this project. I hope that you will all enjoy it.

PREFACE

I had the good fortune to attend the Indianapolis 500 during, what I think, was the most exciting era of racing at that institution. It was an era when you were judged by how fast you went, not by how much sponsorship money you could bring to the team. It was also an era when, if you had the right credential, a photographer could roam the speedway at will. It took me about five minutes, on the first day that I arrived at Indianapolis, to realize that the real pictures were the ones that could be shot from outside the track. This, of course, necessitated special equipment which was hard to come by in those days. Once I was able to purchase what I needed, I rarely came down to the infield where all of the other photographers were stacked up like rows of corn waiting for a crash. I hated the crashes and I hated the people who went out just to photograph someone hitting the wall, or worse. Most of the drivers were my friends and so were their families and I didn't want to see anyone hurt.

Those of us who were there during that time saw something very special. We saw front engine cars, rear engine cars, four-wheel drive cars, and turbine cars. We heard the unforgettable sound of a Novi engine and we heard a car that made no noise at all. We saw many of the great drivers from Grand Prix, NASCAR, sports cars, and drag racing come to Indianapolis to, as the English put it, have a go. We saw the tires progress from big skinny tires to smaller, fatter ones. We saw an era when a three-year-old car (sometimes even older) could still be competitive and, maybe with the right luck, even win the race. We saw great advancements in driver safety, engine development, and chassis development. Most of all, we saw an era that is gone forever and I'm glad that I was there to see it.

Dave Friedman
September 1996

As the pace car brings the starting field through Turn 1, Jack Brabham can be seen well back in the middle of the pack. "I was a bit tense before the race because I had no idea what to expect. Once we started, I found out that it wasn't as bad as I thought it might be. There were no dramas at the start of the race and there was no problem running in traffic with the other cars as there had been in other years. All of the drivers that I competed with in 1961 were very capable professionals and they always conducted themselves as such."—*Jack Brabham*

1961 & 1962
SMALL CAR, BIG NOISE

THE winds of change began blowing at the first U. S. Grand Prix that was held at Sebring, Florida, in December of 1959. Rodger Ward came to Sebring hoping to blow away the Grand Prix cars with his Offenhauser-powered Kurtis dirt track midget. Ward and the Leader Card team for which he drove, believed that the midget's cornering ability would more than compensate for the extra power that the Grand Prix cars had on the straight. Shortly after practice began, Ward found that the Formula One cars could, and did, leave him in the dust. During that race weekend, Ward discussed the possibility of bringing the Cooper Formula One car to Indianapolis with John Cooper and Jack Brabham. It was decided that the Cooper team would come through Indianapolis and test the car, after competing in the October 1960 Formula Libre race at Watkins Glen and before the U. S. Grand Prix, which was to be held at Riverside in November of 1960.

Dr. Frank Falkner and Rodger Ward organized the trip and Ward was there to help Brabham learn his way around the two and a half mile oval. By the time that Brabham had completed his test, he had pushed his lap speed to 144.834 miles per hour, a lap speed that would have put him in eighth qualifying position for the 1960 Indianapolis 500. The secret was in the cornering. Brabham's 2.5-litre Cooper was much faster through the corners than the heavier 4.2-litre Offenhauser powered front engine roadsters, but the little rear engine car ran out of speed on the straight. Once Cooper's team decided to make a serious effort at Indianapolis, things began to change quickly, and when the Cooper team returned to Indianapolis in May of 1961, history was made and the whole face of Indianapolis racing was changed forever.

John Cooper brought his Cooper Climax powered Formula One cars to Sebring for the first ever United States Grand Prix. His team clinched the first world championship ever won by a rear engine car. "Ah, the Cooper Indy project, well I'd better go through that one with you. First of all we had a doctor by the name of Frank Falkner who was the doctor for the entire Grand Prix circuit as it was then. Frank had been a pediatrician and had done a lot of work for the American government on children's health. He always made sure that he got to all of the Grand's Prix and he would always give a lecture at the hospital nearest to the area where the Grand Prix was being staged regarding emergency driver care. He also used to look after all of the minor ailments of all of the Grand Prix people. When we went to Sebring to compete in the first United States Grand Prix and win our first World Championship in December 1959, Rodger Ward, who had won Indianapolis that year, had been entered in a midget. He said, and he really felt, that he was going to blow us all off. Well, it didn't take long for him to realize that it wasn't going to quite work out that way. Rodger really didn't realize what a modern Grand Prix car was like and when everyone got to the first corner, the Grand Prix cars just murdered his midget. Well, he took it quite well and after the race he said to me, quite jokingly I think, 'You should bring one of your cars to Indianapolis.'"—*John Cooper*

Jack Brabham's fourth place at the 1959 USGP assured the win of his first of three world championships.

Rodger Ward made a valiant, but futile attempt to stay competitive with the far more sophisticated Formula One cars of the time. Ward drove the Leader Card midget powered by a small Offenhauser engine. His 'off the cuff' remark to John Cooper suggesting that the Formula One Cooper brought to Indianapolis for a test was not taken lightly.

Jack Brabham is seen here competing in the USGP at Riverside in November 1960. By the time this race was run, Jack had already won his and Cooper's second consecutive world championship. Brabham probably tested this car at Indianapolis in October 1960. Unfortunately, no known photographs exist of the Cooper test at Indianapolis. "We took the Cooper to Indianapolis in October 1960 after several discussions with Rodger Ward who was, at that time, one of the top drivers on the USAC circuit. Our first venture there was a testing session to see what our 2.5-liter Formula One car was capable of doing. I remember turning laps in the neighborhood of 144 to 145 miles per hour and that was with a car, and driver, not really set up to run on that track. Considering that the pole position for the 1960 race was a bit over 146 miles per hour, we were not that far off. If our car had been set up for the unique track conditions that Indianapolis presents, I'm sure that we could have beaten, what was then, the lap record. When we did the test at Indianapolis, it naturally attracted a lot of attention among the drivers and the teams that maintained shops at the speedway. The goodwill and help that was given us was really tremendous and there seemed to be no animosity toward what we were doing in any way. Once we returned to England and the decision was made to run the 1961 Indianapolis 500, things progressed in a rapid way."—Jack Brabham "Frank Falkner was with us and we agreed that we'd dispatch one of our cars to Indianapolis for a test before the final Grand Prix of the 1960 season which was at Riverside, California. Frank and Rodger Ward agreed to organize the Indianapolis trip. When we arrived at Indianapolis we had a lot of American racing people at the track who were curious and wanted to see what an underpowered Grand Prix car would do there. Rodger Ward drove Jack Brabham around the track and gave him some pointers since Jack had never driven on a circuit like that before. Before we could start the test, Jack had to take and complete his rookie test. This was required of all drivers regardless of experience and Jack got pissed off because the officials wouldn't let him go flat out. Instead he had to build his speed gradually until he completed all phases of that test. Bare in mind, we were just running a 2.5-liter rear engine Grand Prix car as compared to the 4.2-liter Offenhauser powered front engine roadsters that normally ran there. Once Jack completed his test, he turned in some pretty respectable times that were not far off the lap record. After we completed the test, everyone said 'Why don't you put a bigger engine in that car and bring it back for the race in 1961?"—*John Cooper*

"I became involved in getting the Cooper to Indianapolis because Jack Brabham and I became very good friends after I drove the Leader Card midget in the Formula One race at Sebring in December 1959. Over the next year or so, we talked from time to time and one day, when we were at a race together (I think it was in Cuba), the subject of Indianapolis came up again. Jack said 'We would certainly like to do that but I don't know what Mr. Cooper's thoughts might be along those lines.' I told him that if they ever had any interest in running there, that I could certainly arrange for them to conduct a test there. Well, a little time went by and one day Jack called me and told me that the Cooper team was coming to Watkins Glen for a Formula Libre race and that, if I could organize it, they would like to come to Indianapolis and run a few laps to see what the car would do there. I organized it and I told him 'Bring the car, it's all arranged.' Well they came to Indianapolis and ran some fairly respectable speeds even though the car was not set up to run there and Jack had never run on anything that remotely resembled that type of circuit. After seeing the positive results of the test session, Cooper decided that Indianapolis was something that they wanted to do."—*Rodger Ward*

Soon after his arrival at Indianapolis in May of 1961, Jack Brabham undergoes a driver briefing from Chief Steward Harlan Fengler.

Having arrived at the track after a last minute rush in England, Brabham begins shaking down the Cooper. "Cooper designer Owen Maddock designed a car that was based on the car that we had won the world championship with in 1960. Changes in the petrol tank size, suspension, and engine placement were some of the things that had to be incorporated into the new car. Dunlop had to come up with a tire that could work on the unique brick surfaced track and that would last for 500 miles and Coventry Climax had to come up with an engine that could produce enough horsepower to make us competitive with the Offenhauser's. As it was, we wound up with an engine that gained 24 horsepower (252 bhp) over the Formula One engine that we had tested with the previous October. We, of course, were running on the methanol based fuel that the Indianapolis cars used at that time. It was our hope that we would get a couple of test days in at Goodwood before departing for the states, but because of the last minute rush, we arrived at Indianapolis with a brand new, untested car."—*Jack Brabham*

Rodger Ward (right foreground) gives Jack Brabham some last minute pointers before the Cooper takes to the track. "In 1961, Cooper returned and their car qualified for and ran in the 500. USAC made a couple of concessions in regard to running the car. I think that they allowed them to run a car 92 inches in length as opposed to 96 inches for us and they also allowed them to run on Dunlop tires which was really no concession at all."—*Rodger Ward*

Eddie Sachs never met an interviewer he didn't love and he was never at a loss for words or stories when he found someone who would listen to them.

During the month of May, many a good four-cylinder Offenhauser engine became a three-cylinder piece of junk in a matter of a few seconds. This photograph illustrates the old racing expression "sawed in half."

One of the roadster chassis undergoes a complete cleaning under the watchful eyes of the many hardcore fans who used to gather daily by the fences of old Gasoline Alley.

On May 12, 1961, one of American racing's most popular figures, Tony Bettenhausen, was killed while testing a car for his old friend Paul Russo who was having handling problems with the car.

Eddie Sachs takes time out to satisfy the Army's autograph requests.

Andy Granatelli (left) is conferring with famed Novi driver Paul Russo. Granatelli had just purchased the Novis and he brought them back to the speedway hoping that he could change the years of bad luck that plagued this popular team.

Rodger Ward awaits the word to start his new A.J. Watson built roadster. "I knew in 1961 when Jack Brabham came to Indianapolis that the rear engine car was the way to go and I also knew that the roadster's days were numbered. I began to talk to my mechanic A.J. Watson about how the car was lighter and cornered better, but it was 1964 before we built a rear engine car. Many people just couldn't see, or didn't want to see, what was fast becoming a reality."—*Rodger Ward*

The beautiful little Cooper sits for its portrait on the front straight as the straight was when it was paved with bricks. "The car that we raced at Indianapolis was not a Grand Prix car, but a special car that was built for the Indianapolis race only. We built it with an offset chassis, special wheels, tires, fuel tanks and so on. If you think about it, three years later (1965) there wasn't a front engine car left that could be competitive in the race. When we came to Indianapolis, we knew that we would be down on power but we knew that the car was lighter and that we could maneuver better in the corners and we hoped that would be the difference. We found out, rather quickly, that Jack could drive the thing flat out through the corners and that's where we made up a lot of time."—*John Cooper*

Jack Brabham prepares to take a practice run as John Cooper stands to the right of the car wearing a Dunlop driver's suit. "Two things that I used for the first time at Indianapolis were a safety harness and a roll over bar, and it was something that eventually became standard in Formula One. We found that the car didn't call for much modification after our first practice session."—*Jack Brabham*

As Jack Brabham sits in the pits watching minor adjustments being made, the Cooper's sponsor, Jim Kimberly (rear of the car), watches the competition. "We couldn't have mounted our Indianapolis effort without the help of Jim Kimberly who sponsored our project that year. Without him, we could have never gotten the car to the states."—Jack Brabham "Frank Falkner knew Jim Kimberly quite well and Kimberly agreed to sponsor our effort at Indianapolis in 1961. Kimberly was a noted sports car enthusiast and he owned a large Kleenex concern. As I remember it, he gave us $25,000 and agreed to pay all of our air bills because we were flying back and forth between Indianapolis and the Grand Prix races at Monaco and Zandvoort, Holland."—*John Cooper*

A determined Jack Brabham qualified the Kimberly Cooper at a speed of 145.144 miles per hour. That was good enough for 13th starting position. Eddie Sachs was the pole sitter with a speed of 147.481 miles per hour. "The people treated us very well during our stay at Indianapolis and I had a marvelous time there with all of the different people that I met. One example of how well the people treated us was when we were qualifying. We had worked all night so we could get qualified early Saturday morning because we had to fly off to Monaco to race the next day. Well Jack went out and ran his four laps for an average of 145.144 which was good enough to make the race. When he came in after qualifying, the officials said that his time didn't count because he hadn't raised his hand to signal that he was going to qualify and he'd have to go again. When this happened, I was absolutely livid and I said 'Forget it, we're off to Monaco, just forget it.' All of this was being said in front of J.C. Agajanian who was one of the real powers in Indianapolis racing. Agajanian said to me 'What's the problem?' I explained our problem to him and he told me to come with him. We went to the officials and Agajanian said that he'd seen Jack put his hand up and that his time should count. After that, we were in and I asked someone 'What the hell had happened' and they said 'No one says no to Agajanian.'—*John Cooper*

After qualifying, a great deal of time is spent by drivers and crews practicing pit stops. Here the crew of Jim Rathmann's Simoniz goes through its paces. I often thought how great it would have been if I could have stood in this position during the race instead of behind the crowded pit wall at the right of the picture. Note the use of air jacks on the Rathmann car. These time saving devices were in use at Indianapolis several years before they appeared on sports cars running in endurance races.

Jim Rathmann (4) Simoniz passes A.J. Shepard (73) Travelon Trailer as Shepard exits the pits. Rathmann finished 30th and Shepard finished 26th.

With John Cooper at the wheel of the car, the pit crew of the Kimberly Cooper practice their assigned race day assignments. The Cooper team did not have an air jack system installed in their car and, therefore, they were forced into making longer pit stops.

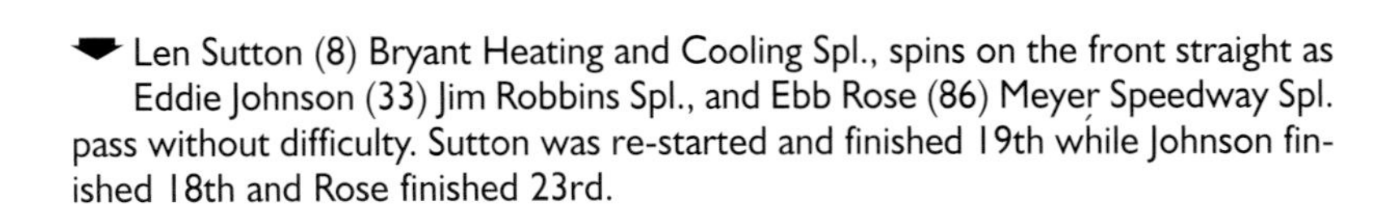

Len Sutton (8) Bryant Heating and Cooling Spl., spins on the front straight as Eddie Johnson (33) Jim Robbins Spl., and Ebb Rose (86) Meyer Speedway Spl. pass without difficulty. Sutton was re-started and finished 19th while Johnson finished 18th and Rose finished 23rd.

The old and the new. Jack Brabham (17) Kimberly Cooper leads Paul Goldsmith (10) Racing Associates through the first turn. Brabham finished 9th and Goldsmith finished 14th. "On race day, the Cooper finished ninth and the whole team really enjoyed the experience. Jack became very excited about Indianapolis and came back several more times. But the Cooper Car Company never returned because their operation was very small and their other commitments prevented them from mounting the necessary effort needed to run at Indianapolis. That car, running the race in 1961, started the rear engine revolution."—*Rodger Ward*

"In 1961, I watched Jack Brabham and the Cooper effort and I compared the Cooper to the outclassed D type Jaguars that ran at Monza against the Indianapolis roadsters in 1957. I was trying to get a feel for what was happening because, by then, the rear engine car had become completely dominant in Formula One. I felt, along with some others, that the front engine roadster was somewhat obsolete, although the roadsters seemed to hang on because of some of the wonderful expertise that Americans seem to have. Seeing all of this that year was what planted the seed in my mind. I could see that Jack Brabham wasn't as serious about the Cooper effort as he might have been if he hadn't been forming his own racing operation at the time. Jack did better at Indianapolis in 1961 than many people realized because, at the time, most people involved with that race had written him off and laughed at him. If you looked below the surface you could see that the Cooper effort wasn't that far off the mark considering everything that was involved. That memory was a rather important factor for the future."—*Dan Gurney*

A.J. Foyt makes one of his pit stops in his Bowes Seal Fast Spl. Foyt went on to win the first of his four Indianapolis 500 wins that day.

A.J. Foyt pulls into victory lane after a dramatic win in which he snatched the lead from Eddie Sachs after Sachs had to make a late race pit stop. With three laps to go, Sachs had to replace the right rear tire because it was shredding. Foyt beat Sachs by 8.28 seconds. After the race Foyt said "God was with us, it was just our day."

The Kimberly Cooper Spl. was forced to make an unexpected third pit stop due to excessive tire wear. The team strategy was to make only two pit stops therefore making up for the car's lack of power on the long straights. "We figured that we could do the race on two pit stops, which would make up for our lack of horsepower and pit stop experience, but Vic Barlow from Dunlop insisted that we would have to make three pit stops. Vic was right because we had to make an unscheduled, our third, pit stop near the end of the race. During the race, I realized the limitations of the Cooper. Although I could go through the corners quicker than the roadsters, I lost that advantage when I got on the straights where the roadster's additional horsepower enabled them to walk away from me. It didn't take long for me to realize that, if I wanted to come back to Indianapolis, I would need at least 100 more horsepower to be competitive. The car went very well during the race and the engine and brakes performed better than I ever hoped for. We finished ninth and we were extremely happy with that result because John Cooper and I had figured that the effort would be a success if we could finish in the top ten. I really enjoyed the challenge of the Indianapolis 500 and, as you know, I returned to run there again on several occasions."—*Jack Brabham*

Brabham is applauded by the competition and officials alike as he pulls into the pits after a fine 9th place finish. "We didn't expect to win but we wanted to put on a good show. I think, after the race, the people at Indianapolis realized that the time had come to put the engine in another place and make the car handle better. If we hadn't had a bad pit stop during the race, we would have finished fourth instead of ninth. The writing was certainly on the wall for everyone to see. I've been asked many times why we didn't return to Indianapolis in 1962. It was two or three things, really, that prevented us from returning to Indianapolis. Basically it was the fact that we had won the Formula One World Championship in 1959 and 1960 with the 2.5 liter car; we were a very small organization employing only 25 or 30 people; and when the FIA changed the Grand Prix engine displacement to 1.5 liters for 1961 we knew that we couldn't repeat as world champions, so we decided to go to Indianapolis. By 1962, we had a new 1.5 liter Grand Prix car to sort out, we were building and racing the Cooper Monaco sports car and we were building and racing Formula Juniors. We were also involved in a big way with B.M.C. on the Mini Cooper. We were using B.M.C. engines in the Mini Cooper, not Ford engines and, at that time, that would have certainly hurt us in any future Indianapolis endeavor because the Ford engine was about to take over in that type of racing. We just didn't have the money, the time, or the staff available to organize the proper effort that it would have taken to do well at Indianapolis that year. Another major problem was that, in early 1962, Jack Brabham had left Cooper to form his own racing concern and Bruce McLaren, my other driver at the time, had very little enthusiasm for that type of racing."—*John Cooper*

A youthful Roger Penske (left) wishes friend and road race rival Dan Gurney good luck before the start of the race. Before long both of these men would make an ever-lasting impact upon the Indianapolis racing scene.

1962
A NEW ARRIVAL

EVEN after Jack Brabham's startling performance in 1961, the regular teams that competed at Indianapolis every year failed to get the message. The only serious rear engine effort that was mounted in 1962 was the one by Mickey Thompson and his team of drag racers. John Zink also appeared that year with a turbine-powered rear engine car, but the engine power left something to be desired and it did not qualify.

A number of changes occurred in 1962. The biggest change was that the old brick speedway had now been paved and only a three foot strip of bricks remained at the start-finish line. Other changes had people like Colin Chapman, Roger Penske, and Zora Duntov checking out the scene and Ford Motor Co. executives were seen making notes.

Well-known all-around driver Dan Gurney also appeared at the speedway for the first time. After qualifying eighth fastest, Gurney drove the only rear engine car in the 1962 race. When he and Colin Chapman left the track for their meeting at Ford, history would be made.

In 1962, another Grand Prix star arrived on the Indianapolis scene. That gentleman's name was Dan Gurney and he would make a major impact on the future of oval track racing. By April 29, Gurney had taken and completed his rookie test in John Zink's roadster. "1962 was the first year that I went to Indianapolis as a driver and I was driving for John Zink. I drove a Watson roadster that was prepared by Denny Moore, Zink's chief mechanic, during my rookie test and then I was supposed to drive a Boeing turbine powered Lotus that, according to their calculations, looked like it had a real chance"—*Dan Gurney*

This is the turbine powered Lotus that Dan Gurney was to drive. "What John Zink had done was stretch the Lotus chassis in the engine compartment in order to make the turbine, which was rated at 350 bhp, fit. That was quite a bit shy of what the better Offys were putting out at the time."—*Dan Gurney*

Gurney leaves Gasoline Alley to make a practice run in the Zink turbine. "I couldn't make the thing run faster than 143 miles per hour and I finally said to John Zink 'If you can find someone who can drive this car faster, get him because I've reached my limit in this car.' The car wasn't bad in the corners but it couldn't run down the straight. A little known fact was that the car arrived late at the speedway because John had tried it out in Oklahoma and had flipped it. Once I found that out, it didn't particularly inspire my confidence in the car's performance. In the week before the first weekend of qualifying, I realized that I wasn't going to get the job done that John expected so I stepped out."—*Dan Gurney*

The three Mickey Thompson entries in 1962 were Buick powered and were designed by John Crosthwaite who had never seen the speedway prior to his arrival with the cars. Note that Fairman's car is sponsored by Jim Kimberly who had sponsored the Cooper effort the year previous. Unfortunately, Fairman ran into various problems on the track and was replaced by a parade of drivers who attempted to qualify the car for the race.

◀ You can't say there weren't some creative minds at Indianapolis in 1962. Jim Rathmann tried this inverted wing on his Simoniz Spl. to see if he could get more downforce through the corners. It didn't work to Rathmann's satisfaction and was discarded. This car, in its conventional form, qualified in 23rd starting position and finished 9th overall. This was also the car in which Rathmann won the 1960 race.

▼ Dick Rathmann (9) Chapman Spl. leads Chuck Daigh (35) Mickey Thompson through the first turn during practice. Daigh ran into numerous mechanical difficulties and failed to qualify while Rathmann qualified in 13th position and finished 24th overall.

Dan Gurney has a conversation with Duane Carter, the driver who replaced him in the John Zink turbine. "Duane Carter, Pancho's dad, took my place in the Zink car. Duane had years of experience at Indianapolis and he couldn't get very much more out of the turbine powered car than I could. He couldn't get the car up to qualifying speed and it got parked. His experience in that car made me realize that the problems I was having with that car were not of my doing and that made me feel better about my driving ability at Indianapolis."—*Dan Gurney*

Dan Gurney gets some technical advice from Mickey Thompson prior to taking his new car out on the track for the first time. "It was fairly late in the game when I made a deal with Mickey Thompson to run one of his Buick powered rear engine cars. I was able to make that deal because Mickey was without a driver for that car. I got in Mickey's car two days before qualifying and I qualified at 147.80 mph, ninth fastest in the field of 33 starters. That car was a great car with a lot of potential but unfortunately the engine was built based on drag racing principles which just didn't work at Indianapolis. In spite of the engine, that car ran pretty damn well."—*Dan Gurney*

Mickey Thompson at work on one of the Buick engines used in his 1962 Indianapolis effort. Unfortunately, these engines were the Achilles heel of those cars and were one of the main reasons that only one of the three cars qualified for the race.

Rodger Ward checks out the fuel tank of his A.J. Watson built Leader Card roadster.

A.J. Foyt and his chief mechanic George Bignotti (background) formed one of the most feared racing combinations in American racing history.

The cars and their crews wait in line for qualifying to begin. Note the photographer at the far left of the photograph holding one of the old 4x5 Speed Graphics. Boy, does that bring back painful memories.

Dan Gurney and Chuck Daigh (right) make some adjustments to the windscreen. Daigh was a very good driver, but he was an even better mechanic.

Parnelli Jones (left) is interviewed under the watchful eyes of his car owner and mentor J.C. Agajanian (center). On the first day of qualifying in 1962, Jones became the first driver to officially break the 150 mile per hour barrier with a average qualifying speed of 150.370 for four laps and a one lap record of 150.729 miles per hour. Brake problems during the race caused Parnelli to finish in seventh position.

Rodger Ward rolls out for practice in his brand new Leader Card 500 Roadster. Ward was the second fastest qualifier for the 1962 race with a average speed of 149.371 miles per hour. In spite of Ward's warning about the rear engine car being the wave of the future, only Mickey Thompson's team showed up with a serious rear engine effort.

Under the watchful eyes of a crowd of officials and photographers, Dan Gurney leaves the pits for his qualifying attempt. Gurney qualified eighth fastest in the field with an average speed of 147.886 miles per hour. This was the only rear engine car to make the race in 1962. "Mickey Thompson brought several rear engine cars to the speedway in 1962 and Dan Gurney drove one of those cars very well. I was amazed that after the Cooper's good showing in 1961, that no other team made a serious rear engine effort in 1962."*—Rodger Ward*

Dan Gurney prepares to make one of his final practice runs before race day. Dan is under the watchful eye of Mickey Thompson (blue uniform) and Jerry Eisert (red shirt). You will note that the car is now blue instead of white as it was when it was qualified. You will also note that the car has picked up the Harvey Aluminum sponsorship.

Dan Gurney, Zora Duntov (center), and Chuck Daigh enjoy a moment during a lull in the action.

A.J. Foyt prepares to defend his race title.

34
Mickey
THOMPSON
HARVEY
Firestone

Mickey Thompson helps strap Dan Gurney into the car prior to the start of the race. Not only was this the only rear engine car in the race, it was powered by the only production based engine (Buick) in the race.

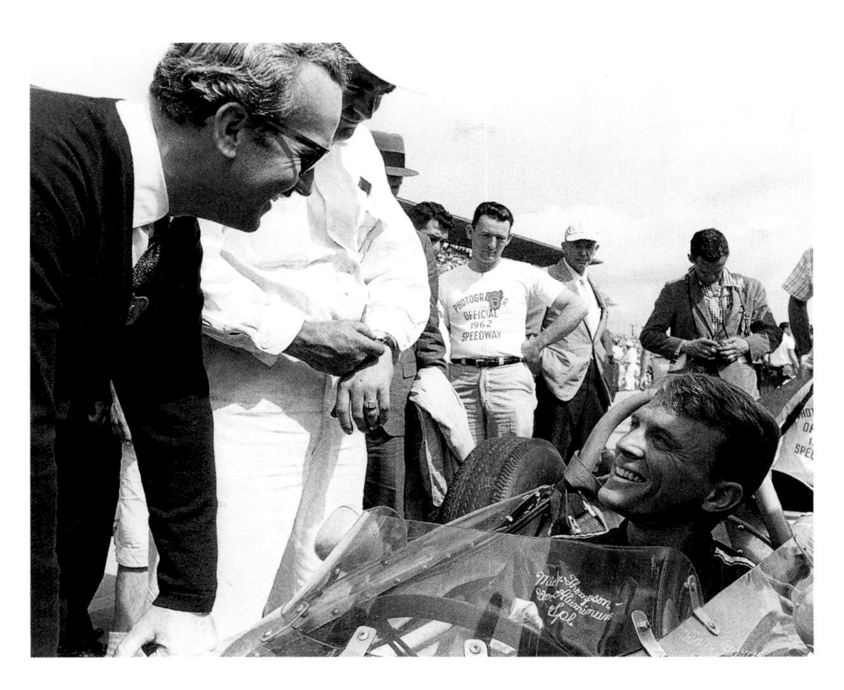

Lotus designer and builder Colin Chapman (left) wishes Gurney well as Mickey Thompson (center) looks on. "After the 1961 race I was convinced that it wouldn't be long before the rear engine car would dominate this race, so why not be in on the ground floor. In all of this process, I could see that there was one main rear engine pioneer in Europe and that was John Cooper and his dad Charles, and their drivers Jack Brabham and Bruce McLaren. However there was one virtuoso, as a design man, and that was Colin Chapman. His first attempt at a Formula One car to combat the Coopers, which were the dominant cars at the time, was the Lotus 18 and I think that it was a better car. It was not better aerodynamically, but it was lighter and more advanced in many ways. I convinced Colin Chapman, after buying him a plane ticket, to come to Indianapolis in 1962 and case the joint, so to speak."—*Dan Gurney*

Dan Gurney (34) Harvey Aluminum Spl. and Bobby Marshman (54) Bryant Heating and Cooling Spl. race down the front straight. Note the huge crowd in the background.

A. J. Foyt's pit crew at work. Foyt's luck ran out in this race when he lost the left front wheel and spun his car on the 70th lap. Foyt's bad fortune illustrates the old racing saying "You can go from the penthouse to the shit house in a real hurry."

This photograph illustrates something never seen in racing these days. Elmer George started the Sarkes Tarzain Spl. from 17th starting position but came in for relief after 59 laps and Paul Russo, whose car was out of the race, relieved him. After 130 laps, A.J. Foyt (seen here getting in car) relieved Paul Russo. On lap 146, Elmer George (the original driver) attempted to get back into the car but the engine seized and the car was out of the race. The car finished in 17th position. Imagine this scenerio today involving three different drivers from three different teams driving one car? Not on your life.

◀ On Lap 94, Gurney coasted to a stop in the first turn due to rear end and engine problems. "I was very proud of my performance in that race and we showed that the car was damn good. The car handled very well and, with a little more development, had a lot of potential. In the race I was in trouble from the beginning even though I was running in the top ten. Finally the engine and rear end let go well before the half way point. It was great fun while it lasted and Mickey was tickled pink after the race."—*Dan Gurney*

▼ Rodger Ward wins the 1962 Indianapolis 500. "The first time that I won the Indianapolis 500 in 1959 (pictured here) I thought that I might have been lucky but when I won again in 1962, I knew that I had really accomplished something."—*Rodger Ward*

Dan Gurney prepares to take to the track during the March 1963 testing session. During the March test, Gurney did a lap at 150.501 miles per hour which was, at this time, the second fastest time ever recorded at the speedway. When the test concluded, the Ford engine had run 457 miles, including 390 miles at racing speed, with little or no trouble.

1963 THE LOTUS INVASION

1963 will always be remembered as one of the most exciting years in the history of the Indianapolis 500. The Lotus Fords appeared for the first time and they were probably the most watched and most exciting cars of the month. The 700bhp Novis returned and all three of their cars made the race for the first time in five years. Mickey Thompson returned with a radical new car sporting low profile 12-inch tires. And Parnelli Jones won the race after a heated debate over whether or not he should have been disqualified for leaking oil.

Another big controversy in 1963 was the debate over tires. The normal size rubber used by the front engine roadsters was 18 inches in the rear and 16 inches in the front. When practice started, the Lotus Fords were using 15-inch tires and Thompson was using 12-inch tires. The front engine people complained that the smaller wheels and tires gave the rear engine cars an unfair advantage because they had a wider tread and were able to put more rubber on the track.

To add fuel to the fire, Goodyear showed up at the track for the first time since 1922 with their 15-inch stock car tires (made for A.J. Foyt) which had a wider track and lower profile than the Firestone tires being used on the Lotus Fords. Finally Firestone made their 15-inch tires available and many of the teams switched to the smaller tires in their search for more speed.

This created a huge problem for the wheel manufacturers (particularly Halibrand) because they were not prepared to supply the smaller wheels in such large numbers or on such short notice. Finally Halibrand was able to supply the necessary wheels and the controversy subsided.

If you were at the speedway in 1963, you were starting to see the makeup of the driver contingent change. Drivers were not only coming from the sprint car and midget ranks as they had for many years, but they were now coming from NASCAR, sports cars, Formula One, and drag racing. New faces were all around that year—Bobby Unser, Johnny Rutherford, Jim Clark, Graham Hill, Pedro Rodriguez, Masten Gregory, Bill Krause, Junior Johnson, Curtis Turner, and Art Malone to name a few. Yes, 1963 was a very exciting year.

The prototype Lotus 29 was rushed to completion so that the car could be flown to the United States for testing at the Ford test track at Kingman, Arizona and at Indianapolis. The four speed Colotti transmission and the 260ci Ford engine are bolted into the chassis at the Lotus factory. The 260ci engine was selected for the project because it had been well proven in the Shelby Cobra. The biggest difference between the Cobra engine and the Indianapolis engine was the fact that the block was made out of aluminum and the engine weighed in at just 350 pounds. "While at Indianapolis in 1962, Colin and I met with some of the Ford people who I knew through my efforts in stock car racing. After the race was over, Colin and I went to Dearborn and met with Bill Gay, Don Frey, Dave Evans, and Bill Innes who was the head of Ford's engine and Foundry Division. In those days, Ford was a bit behind the times as far as engine design went and when we started working on the Indy project, they wanted to copy the Offy engine. The people at Ford actually got a hold of an Offy engine and started copying it. In the meantime, they asked if we could make the Cobra push rod engine work. They asked us how much horsepower would we need to be competitive. After doing as much calculation as we could do in those days, it was figured that we would need a bare minimum of 350bhp, but we wanted at least 365bhp. Pretty soon, we had ourselves a deal with Ford and I was in on the ground floor."—*Dan Gurney*

Jim Clark shakes down the prototype Lotus 29 Ford at Snetterton race circuit in England prior to the car being painted, presented to the press, and shipped to America for testing at Kingman, Arizona and Indianapolis.

The prototype Lotus 29 Ford stands ready for shipping to America and to Ford's test track at Kingman, Arizona.

Once the car arrived at Indianapolis in March of 1963, the crew set to work preparing the engine for the test session. Note the roll bar which was built on to the engine. In other words, every time an engine was changed, the driver got a new roll bar.

Dan Gurney lapping on the Ford Test Track in Kingman, Arizona in early March 1963. "We went to Kingman and ran some good laps and as I remember they were in the 160 to 165 miles per hour range, but we did have some reliability problems with the engine and we weren't too happy about that."—*Dan Gurney*

The oldest car entered for the 1963 Indianapolis 500 arrives at the airport and is carefully lowered to the ground. This Novi chassis was built by Frank Kurtis in 1956 and had led the race that year and again in 1957 before succumbing to the usual Novi bad luck. The car had not qualified for the race since 1958, but was qualified by drag racing star Art Malone on the third day of qualifying for this race. Malone qualified at 148.343, good enough for 23rd starting position.

"When we showed up at the speedway to test the new Lotus Ford, we found that we were in the ball park right away, but we had a lot to learn. We had done some preliminary testing at Kingman, Arizona and Jimmy had done some testing at Snetterton in England. For the testing sessions, we had not developed a permanent exhaust system yet, so we used a BRM-type exhaust system which looked like what the drag racers were using at that time. We were also using carburetors instead of fuel injection, which at one point, had been discussed. During the testing sessions we used both Firestone and Dunlop tires and in this instance the Dunlops were faster."—*Dan Gurney*

Pedro Rodriguez (right) has passed his rookie test and, with the help of chief mechanic Joe Huffaker, removes the rookie stripes from the rear of his car.

Famed chief mechanic Danny Oaks overhauls the engine of the Demler Spl. driven by Paul Goldsmith.

It would be very hard to miss the Day-Glo orange Novi of Jim Hurtubise on the track, or anywhere else for that matter.

Graham Hill made a brief appearance at Indianapolis in 1963 to try one of Mickey Thompson's radical new Chevrolet powered cars. He didn't like it and returned to Europe. "The so called "roller skate" cars that we built for the 1963 race were a pile of shit and should never have been run. We had a steady succession of drivers and no one liked the cars or could drive the cars with any confidence. They were an aerodynamic disaster and they did nothing but crash and blow engines all month."—*A Mickey Thompson crewman*

Dan Gurney (foreground) and Jim Clark are both headed out to practice in their Lotus 29 Fords. Note that Gurney's car is painted in the American racing colors of blue and white, while Clark's car is the well-known British racing green.

The light weight monocoque Lotus 29 chassis undergoes a rebuild in the Lotus Ford garage at Indianapolis. "I think the reason why the English were so far ahead of us in chassis design was because they had to rely on getting good fuel mileage because of the expense of buying fuel in Europe. We never cared much about that and that is what hurt us in the long run. The English worked on developing light weight chassis and small, fuel efficient engines, and that put them way ahead of the game. This also helped them develop a lot of really good racing people."—*Parnelli Jones*

Dan Gurney races down the front straight past the watchful eyes of chief mechanic Bill Fowler (left) and Colin Chapman (right with stop watch).

The Aston Martin powered Cooper driven by Pedro Rodriguez was the same chassis that Jack Brabham drove to ninth place in 1961. Pedro qualified for the race on the first weekend of qualifying, but was bumped from the field by a faster car on the second weekend of qualifying.

◀ Jim Hurtubise (right) discusses the Novi's unique handling qualities, or lack of them, with rookie driver Bobby Unser. Unser would go on to win the Indianapolis 500 three times before he retired from racing.

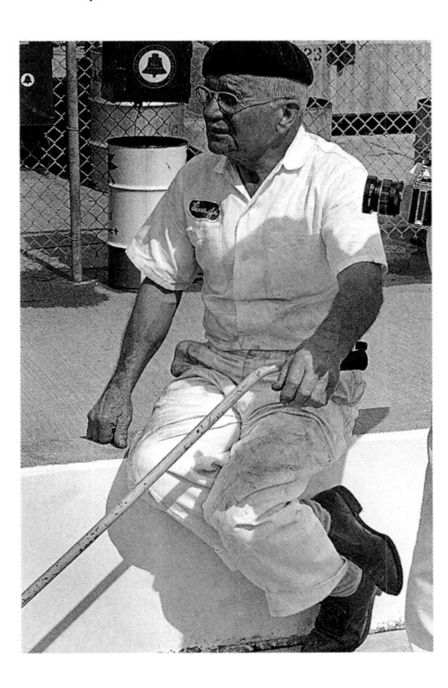

◀ Jean Marcenac, legendary Novi chief mechanic, oversees the preparation of his "children".

▼ The 1963 Leader Card team pose for an informal portrait. Don Branson (5), Rodger Ward (1), and Len Sutton (7) were the drivers. Team owner Bob Wilke is the shorter of the four men in the back row and chief mechanic A.J. Watson is the tallest.

Dan Gurney appears deep in thought while Colin Chapman (left) and Jim Clark (right) stand by for comments. Chief mechanics Jim Endruweit and Bill Fowler are behind Chapman.

The slim body shape of the Lotus 29 Ford is evident when viewed from above.

Parnelli Jones (left) and J.C. Agajanian were one of the most successful owner-driver combinations during this era. "Aggie was like a father to me and he was responsible for most of the good things that happened during my racing career. I was offered rides at Indianapolis two or three years before I finally went there, but I wanted to wait until I had the right equipment and Aggie provided that."—*Parnelli Jones*

Jim Clark prepares for a practice run. "The Lotus Fords really did open our eyes so that we could see that there was certainly another way to go. That effort really did open all of our eyes."—*Parnelli Jones*
Jim Clark was up over the 150 mile per hour barrier during practice, but he was taking his time getting used to this type of racing. Clark was shuttling back and forth between Indianapolis and Europe where he won the British Grand Prix. "In 1963, things really got serious at Indianapolis with the arrival of the Lotus Ford effort. Colin Chapman had Jimmy Clark and Dan Gurney as drivers and they were really very fine drivers. Those damn cars really opened up some eyes at Indianapolis."—*Rodger Ward*

Parnelli Jones' roadster "Ole Calhoun" was appearing in its fourth Indianapolis 500 since being built in 1960 by A.J. Watson. Jones' unofficial lap record of 152.027 miles per hour on May 11 sealed the fate of the old style wheels and tires. Jones' record speed was recorded while using the new Firestone 15-inch tires and it was the fastest lap time recorded all month. Jones attributed his stunning speed to two factors: the new 15-inch tires and a new fuel mixture. Jones claimed that he was running pure alcohol without any nitro and he also claimed that he was getting an extra 500rpm coming out of the corners and that the car felt really secure with the new tires.

Jim Clark, deep in thought. "Jim Clark was a great racer, nice gentleman, and a good friend."—*Parnelli Jones*

Drag Racing star, Art Malone, drove the oldest of the Novis.

Troy Ruttman was one of the greatest ever natural talents to appear in American racing. He was also the youngest man (age 22) to ever win the Indianapolis 500, which he did in 1952. Ruttman was not a great qualifier, but when it came to the race, he would be challenging the leaders within a few laps. "In 1958, I was already embarked on a path that would lead to my becoming a professional race driver. One of the things that happened during that period was that I befriended Troy Ruttman and that friendship has lasted to this day. Troy was one of my heroes when I was coming up and, although we were about the same age, he blossomed a whole lot sooner than I did. We went to Europe together in 1958 and we drove across the continent together. I believe, at one point, that we drove from Monza to the Nurburgring where Troy and I had both managed to put together a ride. During that time I talked to Troy about Indianapolis because, if I was going to be a professional race driver, I was going to go to Indy eventually. That was a foregone conclusion. I really spent a lot of time talking to Troy about Indy and what it took to run there."—*Dan Gurney*

Dan Gurney confers with Colin Chapman prior to going out for morning practice on the first day of qualifying.

Gurney at speed on the morning of the first day of qualifying. Within minutes, Dan was in big trouble.

"When we got to Indy for the month of May, things went pretty well until I hit the wall on the morning of first day of qualifying. I felt that it wasn't a real good way to start the qualifying process. The crash was my fault although there were a lot of extenuating circumstances. The biggest problem we had was that we had only one set of wheels that didn't have any cracks in them. I was running Jimmy's wheels (the uncracked set) and I just tried too hard and lost it. The idea of going backwards through the first turn at Indy, not knowing what to do, was an interesting experience to say the least. Had I had experience at that sort of thing, I might have been able to do something to save it. When that crash happened it put the whole Lotus effort into a deficit which, I think, was a natural reaction. Anyway, Lotus rolled out the spare car that I'm not sure had ever been run and I got in it and went out to qualify. It was very late in the day and I ran a couple of respectable laps but then I got my foot tangled up in the so called safety strap, which was no fun at all, and I aborted the qualifying attempt. It was too late in the day to make another attempt so I qualified on the second day (Sunday) at 149.019 miles per hour which was good enough for the 12th starting position. I blame myself for most of my first qualifying day problems, but there were definitely some other circumstances." *—Dan Gurney*

Parnelli Jones claimed the pole position for the race with an average qualifying speed of 151.153 miles per hour. His fastest lap was 151.847 miles per hour. Both speeds were track records.

Jim Clark was the fastest qualifier of the Lotus Fords. Clark qualified in fifth position with an average speed of 149.750 miles per hour and a fastest lap of 150.025 miles per hour. "I didn't have any big feeling about the Lotus Fords when they came to Indianapolis in 1963. They were there and I thought that was great but, as quick as they were, I was quicker in my roadster. I was really surprised because I thought that they would be much quicker then they were." *—Parnelli Jones*

In the eyes of the record first qualifying day crowd of 220,000, Jim Hurtubise was the real star of the of the day. It had been five long years since a Novi had qualified for the race and Hurtubise put the car in the middle of the second row with an average speed of 150.257 miles per hour and a fast lap of 151.261 miles per hour.

To the delight of the massive race day crowd, Jim Hurtubise (56) took the lead from Parnelli Jones (98) as the field entered the third turn. Hutubise had a three car length lead as he and Jones completed the first lap. Hutubise won a personal bet with Jones over who would lead the first lap. Bobby Marshman (5), Don Branson (4), Jim Mc Elreath (8), and Rodger Ward (1) trail Hurtubise and Jones.

By the end of the second lap, Jones (98) had taken the lead from Hurtubise (56), Marshman (5), and Mc Elreath (8). It was a lead that Jones would seldom relinquish during the race.

Bobby Marshman leads Duane Carter (83), and Lloyd Ruby (52) down the short chute. Of the five Thompson cars at the speedway, only two qualified. "It was a case of too many cars and too few crew members. It was also a major problem that the small wheel cars were just no good. If we had concentrated on the two 1962 cars that we had back there, things might have been different." *—A Thompson crewman* As a matter of record, Eddie Johnson finished ninth in one of the 1962 cars powered by a stock block Chevrolet engine.

Dan Gurney (93) and Jim Clark (92) run together early in the race. Eddie Sachs (9) and Roger Mc Clusky (14) trail the Lotus Fords. "Before the race, I said to Bill Gay 'My engine doesn't feel right, something's wrong with it.' Gay had the mechanics fire it up and he proceeded to rev it up and float all of the valves and everything else. He said 'It sounds fine to me' and that was it. I never had a decent engine and, in the race, I ran on seven cylinders from the word go. Jimmy followed me for awhile, while he was getting his feet on the ground. I remember thinking 'Why in the hell doesn't he pass me on the straight away?' Pretty soon he shot by me and was gone. Of course you had all of the British vs. Americans, front engine vs. rear engine, and Ford money coming in to beat up on the poor Offy business going on. It was a lot of bullshit but it was great copy for the fans and it created a hell of a lot of interest in the race for the general public. However, if you are in the middle of all of that nonsense, it can be rather distracting."*—Dan Gurney*

Bud Tingelstad (54) leads Johnny Boyd (23) down the short chute.

Parnelli Jones makes one of his three pit stops.

Competition is very close in the third turn with Dick Rathmann (8) leading Duane Carter (83), Allen Crowe (35), Eddie Johnson (88), Bob Vieth (86), Art Malone (75), Bobby Grim (26), and Jim Rathmann (16).

The Lotus Fords, looking lost in the pack, prepare to pounce on Jim McElreath (8), Bobby Grim (26), and Don Branson (4). Paul Goldsmith (99) and Ebb Rose (32) trail Clark and Gurney. "That race was a turning point in the history of Indy car racing and the handwriting was really on the wall for the front engine cars to see. At that point in time most of the front engine cars were relegated to the speedway museum. Anyone who was watching what was going on in the racing world at that time, should have realized that what happened at Indianapolis in 1963 was inevitable."—*Dan Gurney*

Parnelli Jones' last pit stop. Look closely and you can see the oil that has blown back on the rear end of the car. The final laps of the race created a huge debate on whether the USAC officials should black flag the leading car for leaking oil. "When we remember the rather controversial finish of that race, we must first remember that both Parnelli and Jimmy were really great drivers. It's true that Parnelli was leaking oil but so what. The USAC officials weren't about to pull him in and Chapman and Clark wouldn't have wanted to win the race that way anyhow. If the race had have been held in England, I'm sure that it would have gone the other way. As it was, Chapman did complain but so what, the race was run and that's the way life is."—*Dan Gurney*

"Many people thought that Jimmy should have won the race because Parnelli's car should have been black flagged for dropping oil on the race track. People also said the only reason that Parnelli wasn't black flagged was because of Aggie (J.C. Agajanian). It is absolutely certain that the car had an oil leak because the oil tank had developed a crack in it. The problem was that by the time USAC had determined who it was that was losing the oil, the oil was below the level of the crack and there was no more leak. Anyway, USAC didn't black flag him and I believe it was the right decision because you can't penalize someone for what they did 10 or 20 laps ago."—*Rodger Ward*

Dan Gurney congratulates Jim Clark on a fine second place finish. "Jimmy was my very close friend and it took me a long time to get over his death. In fact, I almost retired when I heard about it."—*Dan Gurney*

Dan Gurney (12) and Rodger Ward battle for second place early in the race. "When the race started, as you will recall, we had that terrible accident and they stopped the race after just a couple of laps. Under some conditions if there is a serious accident like that and they don't stop the race it's easier to, not put it out of your mind, but put it aside. If they actually stop the race and you have an hour and 45 minutes to sit and think about the fact that two of your friends are dead, it can create a problem. When they restarted the race, I drove a good race mechanically but I didn't have the same desire and the same push that I had before the race started."—*Rodger Ward*

1964
REALITY AND TRAGEDY

THE 1964 Indianapolis 500 will always be remembered for the senseless tragedy that occurred on the second lap when two drivers, Dave Mac Donald and Eddie Sachs, were killed in the worst accident ever seen at the speedway. I will not dwell on my feelings regarding this incident because it still angers me to discuss it even though it happened 32 years ago.

There are many positive things that we should remember about the 1964 race, but don't. May of 1964 brought record crowds to the speedway probably because of all the controversy created by the previous year's finish. Many fan clubs of both the rear engine and front engine cars were noticeable with home-made signs and T-shirts.

A number of rear engine cars appeared from well-known builders such as Watson, Brabham, Vollstedt, Huffaker, Brabham, Gerhardt and Halibrand. Lotus and Mickey Thompson both returned for another try and the roadster contingent was out in full force. A number of new innovations were seen on the roadsters including a four-wheel drive Ferguson chassis for the Novi.

One of the most exciting new innovations seen at the speedway in 1964 was the advent of the new Ford double overhead cam engine. By the time that engine hit the track, everyone knew that it would be a serious contender to unseat the 'tried and true' Offenhauser engine that had dominated the speedway for so many years.

Many of the top Grand Prix drivers also returned to Indianapolis including Jack Brabham, Dan Gurney, Jim Clark, Pedro Rodriguez, Masten Gregory, and Walt Hansgen.

It is interesting to note that both Parnelli Jones and A.J. Foyt had rear engine cars available to them, but rejected them for the older, more reliable front engine roadsters. It is also interesting to note that it must have been the right choice because Foyt won the race.

The Ferguson four-wheel drive Novi chassis is readied for shipping to America. It was thought that the increased traction of the four-wheel drive concept would help insure the Novi's chances of victory.

The howl of the 700+ bhp, supercharged, eight cylinder Novi will never be forgotten by those who had the good fortune to hear it.

SIX OF THE COMPLETELY DIFFERENT CARS THAT APPEARED AT INDIANAPOLIS IN 1964.

Lotus 34 Ford of Jim Clark.

Watson Roadster of Parnelli Jones.

Ferguson Novi of Bobby Unser.

Mickey Thompson Spl. of Dave Mac Donald.

The Watson Leader Card Spl. of Rodger Ward.

Kurtis Novi of Jim McElreath.

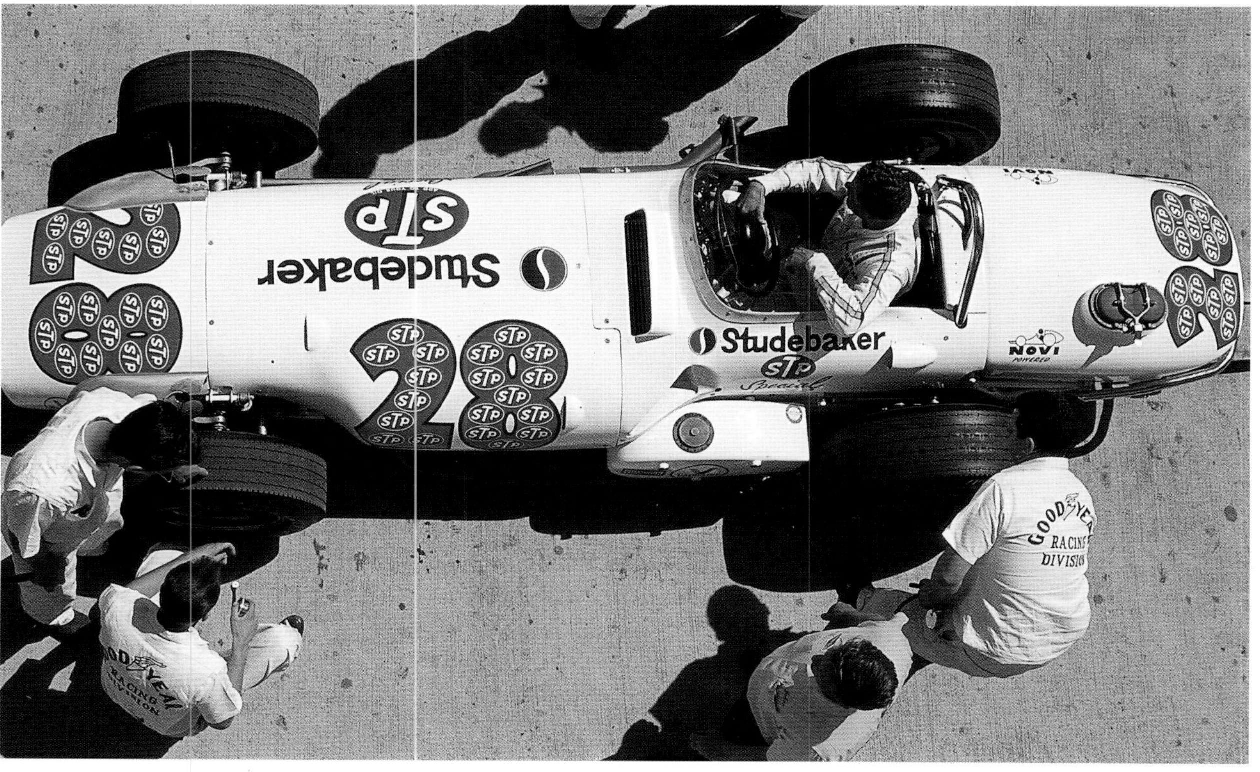

Famed chief mechanic George Bignotti works his magic on one of the Offenhauser engines being built for A.J. Foyt. Foyt and Bignotti proved to be an almost unbeatable team winning 27 races between 1960 and 1965.

Jack Brabham returned to Indianapolis with his own car powered by an Offenhauser engine. Jack was the first qualifier on the third day of qualifying. He had an average speed of 152.504 miles per hour and started in 25th position. He went out of the race on the 77th lap with a split fuel tank, and finished in 20th position. "I remember when Brabham was at the speedway in 1961 with the Cooper. He ran a lot of practice laps then and he used to kid me by telling me 'If they would let me run in the other direction I could go a hellva' lot faster. I said 'Ya, good luck'.—*Rodger Ward*

Dan Gurney and Colin Chapman in serious discussion. Could they be discussing tires?

Sometimes you find rest in the strangest places.

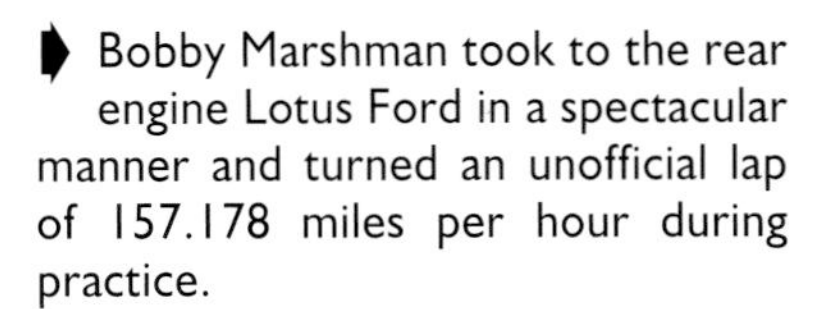

Bobby Marshman took to the rear engine Lotus Ford in a spectacular manner and turned an unofficial lap of 157.178 miles per hour during practice.

Dave Mac Donald testing the aerodynamics of the Mickey Thompson car. He, and everyone else, found these cars to be terrible. Jim Clark followed Dave for several laps on carburetion day and told him to get out of the car immediately and don't drive it on race day. Unfortunately for all, Dave did not heed Clark's advice.

Eddie Sachs had his first rear engine ride at Indianapolis in a Halibrand Shrike powered by Ford. Sachs, known as the crown prince of racing, qualified on the second day of qualifying with an average speed of 151.439 miles per hour. Starting in the 17th starting position, Sachs was killed in the second lap accident that also took the life of Dave Mac Donald.

Rodger Ward finally appeared with the rear engine car that he had tried to persuade A.J. Watson to build several years earlier. "In 1960, I had a chance to take a few laps in Brabham's Cooper when it was at Indianapolis testing. After running that car, even for just a few laps, it was obvious to me that it had some tremendous advantages over the roadsters and that it was the wave of the future. I began to think, right then, that was the way that we were going to have to go. By 1964, I finally convinced A.J. Watson to build a rear engine car for our Leader Card team. We had been present when Rolla Vollstedt was conducting tire tests at Indianapolis in early 1964, and it was there that Watson finally decided to build a rear engine car. The rear engine car that Watson built for the 1964 race was a damn good race car and we deserved to win the race."—*Rodger Ward*

Jim Clark and his Lotus 34 Ford prepare to make a practice run.

On May 7, 1964 Parnelli Jones unofficially broke the one lap record with a speed of 156.223 miles per hour. Parnelli would qualify for the race with an average speed of 155.099 miles per hour and start in fourth position behind three rear engine cars. The protruding air scoop on the nose of the car helps ram more air directly into the intake nozzles of the Offenhauser engine. One of the ways that Parnelli's crew was able to save weight on this car was to use a fiberglass fuel tank that was smoothed and painted to resemble the original metal body paneling. "In 1964 I had my choice of a rear engine car built by Halibrand or my roadster. I chose the roadster because the rear engine car didn't handle that well and my roadster was much quicker."—*Parnelli Jones*

Parnelli Jones is ready to test his new rear engine Offenhauser powered Halibrand car. After several laps, Jones parked it in favor of his tried and true roadster.

A.J. Foyt also had a rear engine car available to him. The car was an Offenhauser powered Huffaker built chassis. Foyt rejected the car and decided to run his old roadster. The Huffaker car was then given to Bob Veith who qualified it on the third day of qualifying at an average speed of 153.381 miles per hour. Vieth started in 22nd position and was up to third place at the end of 70 laps. He went out of the race with engine trouble on lap 88 and finished in 19th position.

A. J Foyt qualified on the first day of qualifying with an average speed of 154.672 miles per hour. Foyt started in fifth position and took the lead for good on the 55th lap. He averaged 147.350 miles per hour for the race and set a new 500 mile race record. This was Foyt's second Indianapolis win in four years. Foyt decided to stay with the car he knew best, his A. J. Watson built roadster. It proved to be the right choice.

Bobby Unser drove a Novi powered Ferguson four wheel drive car in 1964. The car was a radical departure from the Kurtis chassis and Granatelli felt that it eliminated the need to place the engine in the rear. The Ferguson chassis had four-wheel drive, independent four-wheel suspension, and was of a semi-monocoque design. Granatelli felt that such a chassis was ideally suited to the high horsepower Novi engine. He also felt that the Ferguson Novi combination could go much deeper into the corners than the other cars, save on tire wear, and power out of the corners much faster than any of the other cars. Unfortunately, Unser was never able to prove Granatelli's theory because he got caught up in the second lap crash.

Art Malone drove the more conventional Kurtis Novi. Malone qualified on the final day of qualifying with an average speed of 151.222 miles per hour. He started in 30th position and finished 11th. Standing next to the wall in the short chute and having a screaming Novi pass within several feet of my head is an experience that I will never forget, and neither will my ears.

In 1964, Jim Hurtubise drove a roadster designed by himself and his brother Pete. The car featured coil springs instead of torsion bars and was sponsored by, of all businesses, Tombstone Life Insurance. Hurtubise qualified on the first day at an average speed of 152.542 miles per hour, good enough for 11th starting position. He was running third in the race when he lost oil pressure on the 141st lap. Jim wound up finishing 14th.

Dave Mac Donald leaves the pit area to try and sort out the Thompson car to which he was assigned. Note how the wheel wells have been cut away to accommodate bigger tires. Nothing helped the handling of these cars. In fact, they got worse as the month progressed. Numerous knowledgeable racing people advised Dave to step out of the car but, because of his contractual agreement, he wouldn't do it and it cost him his life. "In 1964, those cars should have been withdrawn after the first few days of practice. Because of the new 15-inch tire rule, we were forced to run with the larger wheels and tires and this situation made our bad handling even worse. We had a revolving door of drivers all month and as soon as one driver would get in the car, another would get out. Mickey knew the problems but his ego got in the way."—*A former Thompson crewman*

Signs seen around the speedway during the month of May.

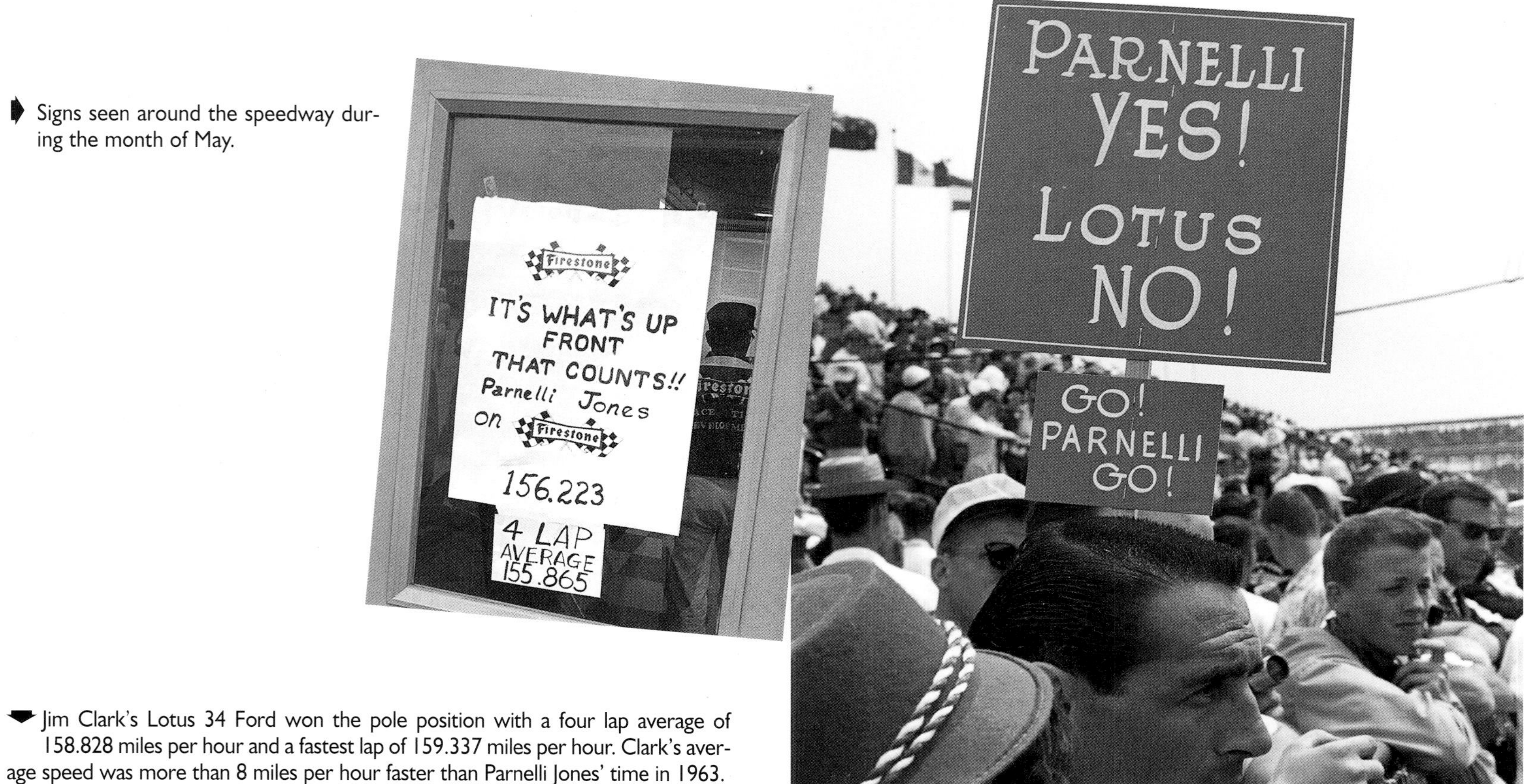

Jim Clark's Lotus 34 Ford won the pole position with a four lap average of 158.828 miles per hour and a fastest lap of 159.337 miles per hour. Clark's average speed was more than 8 miles per hour faster than Parnelli Jones' time in 1963.

The work in the garages never stops. Vince Granatelli (left) and another crew member make adjustments on the Ferguson Novi of Bobby Unser.

Preparation of Jim Clark's Lotus 34 Ford continues as Jim Clark (background) watches. "As far as the preparation of the Lotus cars went, it was OK. Did I feel that the Lotus philosophy of how you do things was very good? No. The car was a product of necessity. Chapman was a brilliant designer and the Lotus people had become used to the inherent risks of a Chapman design so it was like an every day occurrence when problems arose. We had some structural failures in those cars but it seemed, at that time, that it was the price you paid for getting something that was significantly better."—*Dan Gurney*

A.J. Watson burns the midnight oil in the Leader Card team garage. "We screwed up primarily because Ford Motor Company owned the engines and they determined that, on race day, we were going to run on gasoline. I had made up my mind that that was not practical for us. The first reason was that I didn't want to run gasoline because of the fire hazard and the second reason was that my car was about 200 pounds heavier than the Lotuses. Unfortunately, we learned what a fire hazard gasoline was at the speedway by the accident that killed Dave Mac Donald and Eddie Sachs. My car also didn't have the aerodynamic superiority that the Lotuses did and I determined that it would be almost impossible for me to compete evenly with them if we were both running on gasoline. We attempted to set the car up so we could run on 80-10-10 which would give us a little enhancement in the mileage category but would give us the horsepower that we needed to be competitive. Basically I made most of those type of decisions. Every time we went to tune the engine the Ford engineers, who were always hovering about, could tell by the sound of the engine and the smell of the fuel mixture that we were not running gasoline. They actually came into our shop and physically removed the nozzles out of the jets and said 'Rodger you will run on gasoline.' On carburetion day I ran on gasoline and if I'd been half smart I would have continued to run on gasoline because we didn't lose that much by running on it. I thought we were going to lose a lot more than we actually did. The morning before the race I had a meeting with Watson and he said 'Well Rodger, what do you want to do?' I said 'Well, I think we ought to put the shit in.' Watson said 'You know we're not carbureted for that.' I said 'Well, damn it, can't you get close?' and Watson said 'Well we should be able to get close and we do have a three position mixture control valve that we could do something with. Maybe that would help and you can adjust the mixture as the race progresses.' I said 'Great, let's do it.' The crew put the valve in the car and on race day morning Watson said 'Get in the car and see what you think.' I got in and I found that the valve was right on my leg. I told them that I thought in that position, the valve would cut off circulation to my leg. I asked if we could we move it? Well, as you must know, all race drivers are not space scientists, and so in an effort to make absolutely certain that I knew which way was lean and which was rich, I put on the instrument panel "Forward Rich, Back Lean." A bit later, Watson told me to get in the car and see if the position of the valve was OK. I got in and it was fine. Later I asked him what he had done and he told me that he had simply turned the valve over. Dumbshit Rodger Ward didn't realize that when you turn the valve over that it made the instructions that I had written on the instrument panel backwards. I just wasn't thinking and I forgot which position that the valve was supposed to be in."—*Rodger Ward*

Dan Gurney prepares to start the race from the outside of the second row. Gurney qualified with a four lap average of 154.487 miles per hour and was running third when he had to make a pit stop on the 14th lap. "The Lotus crew was top flight and they were on a par with any crew in the racing world at that time. They were all racing for their nation and for their marque and, understandably, they had a great deal of respect for Jimmy. I felt that the crew treated me very well and I never felt that they gave me any less than they gave Jimmy. We were all rank amateurs compared to some of the others at Indianapolis, but we were bringing a lot of new technology to the speedway, which at that time, none of us including Chapman really understood that well."—*Dan Gurney*

Jack Brabham (52) leads Bob Wente (68), Len Sutton (66), Art Malone (3), and Bob Harkey (4) through the first corner. One can see here how much larger the older roadsters look as compared to the newer, smaller, more agile rear engine cars.

Jim Clark makes the long walk back to the pits from the first turn. One fan was heard to tell Clark how shocked he was when Clark's suspension collapsed. Clark responded "You should have been where I was mate."

Rodger on his way to a fine second place finish. "One thing that was a great concern to most of us was that the rear engine car might be a dangerous car to drive because it was so lightly built. We were concerned and we called them "funny cars" because they were not built as rugged as the roadsters. Well, the truth of the matter was that those cars were safer than the roadsters. It was very hard for those of us who had grown up driving the roadsters to jump into those flimsy little cars and go out there, go fast, and feel comfortable doing it. In 1964, it was obvious from the beginning that the rear engine cars were substantially better and faster than the roadsters. The Lotuses of Clark and Marshman ran off and hid from the rest of us while they were running. There was absolutely nothing in that race that could stay with those two cars. Under normal circumstances there was no way in hell that I could have beaten those two, but I was the best of the rest of the field with no problem."—*Rodger Ward*

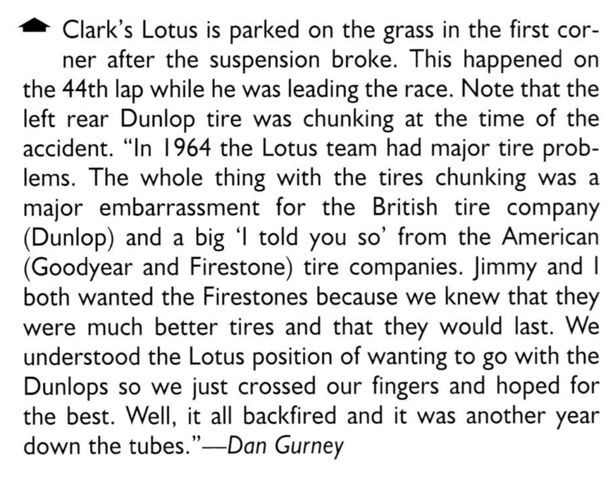

Clark's Lotus is parked on the grass in the first corner after the suspension broke. This happened on the 44th lap while he was leading the race. Note that the left rear Dunlop tire was chunking at the time of the accident. "In 1964 the Lotus team had major tire problems. The whole thing with the tires chunking was a major embarrassment for the British tire company (Dunlop) and a big 'I told you so' from the American (Goodyear and Firestone) tire companies. Jimmy and I both wanted the Firestones because we knew that they were much better tires and that they would last. We understood the Lotus position of wanting to go with the Dunlops so we just crossed our fingers and hoped for the best. Well, it all backfired and it was another year down the tubes."—*Dan Gurney*

⬆ Due to the fact that the American pit crew badly bungled one of Dan Gurney's pit stops in 1963, Lotus decided to use an all English pit crew in 1964.

When Gurney pitted on lap 96 while running in sixth position, the platform jack failed to function properly and the crew had to use the manual jacks to change the tires. While inspecting the tires after the pit stop was completed, it was noted that strips of tread were missing from the center of the tire. After consulting with the Dunlop engineers, Chapman immediately called Gurney in and retired the car after 110 laps.

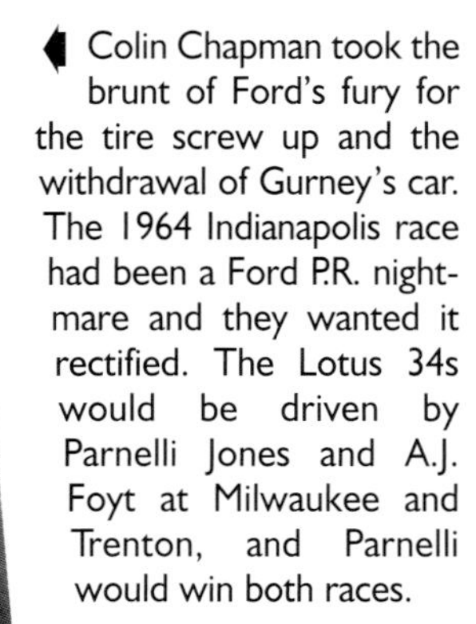

Colin Chapman took the brunt of Ford's fury for the tire screw up and the withdrawal of Gurney's car. The 1964 Indianapolis race had been a Ford P.R. nightmare and they wanted it rectified. The Lotus 34s would be driven by Parnelli Jones and A.J. Foyt at Milwaukee and Trenton, and Parnelli would win both races.

J.C. Agajanian supervises his very efficient pit crew during Parnelli Jones' first pit stop. As Parnelli pulled away from this stop on the 55th lap, his fuel tank caught fire and he had to bail out of the car. Parnelli was given 23rd finishing position.

▲ A.J. Watson (holding left fuel nozzle) and crew service Rodger Ward late in the race. "I made five pit stops that day and they were all for fuel and Foyt (the winner) made two stops. I spent three and a half minutes stopped in the pits and Foyt spent one. I finished second to Foyt, losing the race by less than a minute."—*Rodger Ward*

▶ A grim looking A.J. Foyt has a hard time celebrating his second Indianapolis win after the earlier events of the day. Foyt's wife Lucy sits next to him and George Bignotti sits behind the driver of the pace car.

The first of a long line of Eric Broadley designed Lolas appeared at Indianapolis in 1965. This Lola was driven by Bud Tingelstad. This car was running fifth in the race when a broken wheel put Tingelstad in the wall of turn three on lap 116.

1965 THE LOTUS FORD REDEMPTION

1965 will be remembered for a number of things. Most importantly the race was one of the safest and fastest of the 48 previous races and millions watched all of the drama unfold on closed circuit television across the country.

A number of safety changes were implemented after the 1964 disaster. Stronger fuel cells and smaller fuel loads were just two of the new regulations. Interestingly enough, USAC did not restrict the type of fuel that could be used. A more stringent rookie orientation test was initiated and pressure fueling was eliminated.

One third of the field (11 drivers) were rookies and five of them were still running at the end of the race. Mario Andretti was the "Rookie Of The Year." Other future stars like Al Unser, Gordon Johncock, Walt Hansgen, George Snider, Ronnie Duman, Masten Gregory, Jerry Grant, Bobby Johns, and Joe Leonard would all make their mark in future years.

The number of new rear engine chassis that appeared at the speedway in 1965 proved, beyond a shadow of a doubt, that the roadster was a thing of the past.

There are three facts that will always be associated with the 1965 Indianapolis 500. This was the first race won by a rear engine car. The Ford dual overhead cam engine won its first 500 in just its second year of existence. And Jim Clark became the first man to win the Indianapolis 500 and the Grand Prix World Championship in the same year.

Al Unser was one of the crop of spectacular rookies to arrive at the speedway in 1965. Unser took his rookie test in this Arciero Bros. Maserati powered rear engine car but stepped out when the car failed to get to competitive speeds.

THREE VERY DIFFERENT 1965 CARS AS SHOT FROM ABOVE THE TIRE CHECK LEADING FROM GASOLINE ALLEY TO THE SPEEDWAY.

This car was built by Brawner, Andretti, Moore, and McGee for Mario Andretti to drive in his first Indianapolis 500.

The Ferguson Novi, with Bobby Unser at the wheel, returned for another try at the winner's circle. By now a Novi, in any combination, was hopelessly outclassed.

A.J. Foyt's Lotus 34 Ford certainly had the best color scheme of any Lotus that ever appeared at the speedway.

Al Miller drove the ex-Gurney, ex-Marshman Lotus 29 to a fine fourth place finish in the race. The car was never further back than sixth place, but had to make five pit stops because the fuel tank was only capable of holding 34 gallons.

The MG Liquid Suspension car was built by Joe Huffaker and driven by Walt Hansgen. The car featured a suspension system that was standard on the MG Sports Sedan and it was felt that this system would eliminate a lot of the tire wear that was occurring at that time. Hansgen started in 21st position and was running fifth when fuel line problems caused him to make five pit stops before retiring from the race.

▲ Mickey Thompson's entry for the 1965 Indianapolis 500 was a front engine, front-wheel drive car that was powered by a dual overhead cam Chevrolet engine. Bob Mathouser blew the engine during qualifying and did not qualify. Mathouser had turned laps of 157+miles per hour in practice and that would have put him in the race.

▲ Lotus team Chief Mechanic Dave Lazenby (bottom) and one of his helpers enjoy a quick nap by the pit wall during a lull in the action.

▲ J.C. Agajanian (left), Parnelli Jones, and Johnny Poulsen (right) formed one of USAC's most potent racing teams.

Walt Hansgen prepares to take to the track in his Offenhauser powered Huffaker.

Bobby Unser's Ferguson Novi is pushed off for practice by a crew member wearing what was considered the most gaudy uniform at Indianapolis.

Roger McClusky drove one of two Halibrand Shrikes entered by Dan Gurney's All American Racers while Joe Leonard drove the other. Neither car lasted beyond the 27th lap.

Bobby Unser practices the new Ferguson Novi. Bobby wrecked this car and had to qualify the older Ferguson that he had driven in 1964. Unser qualified with an average speed of 157.467 miles per hour and started in eighth position and was up to fifth before dropping out of the race with a broken oil line.

◀ Jim Hurtubise demolished his Tombstone Life Spl. when the throttle stuck and he slammed into the fourth turn wall during practice on the first day of qualifying. Hurtubise was unhurt.

▶ Jim Clark confers with Colin Chapman before practice. The new Lotus 38s and the Lolas were having problems with broken hub carriers, but Clark was still running laps over 160 miles per hour in practice with no problems.

NASCAR driver Bobby Johns was chosen to be the second driver for the Team Lotus effort after Dan Gurney decided to run his own team and Parnelli Jones decided to stay with Agajanian.

Rookie driver Mario Andretti prepares to take to the track under the watchful eyes of Jim Clark. Mario qualified with a record four lap average speed of 158.849 miles per hour and a record one lap speed of 159.405 miles per hour. These records were later broken and Mario wound up starting from fourth starting position. Mario finished third and was voted Rookie Of The Year.

Rodger Ward missed his first Indianapolis race in 15 years when he crashed his Watson Ford on the third day of qualifying and was 0.151 miles per hour too slow on the fourth day of qualifying. Ward never felt safe in the car and had a hard time getting it up to speed.

Dan Gurney watches as his All American Racers' crew make some final adjustments to his Lotus 38. "In 1965, my racing crew from All American Racers prepared my Lotus Ford for the Indianapolis race. Looking back it was a mistake to take this route because the line of communication with the factory Lotus team wasn't as good as it could have been for obvious reasons. We got along Okay but there was a rivalry between the two teams and it was hard for us to obtain certain parts that were on Jimmy's car. A good example of this was that I didn't have an anti roll bar that was large enough, but Jimmy did and so did Bobby Johns who was the second Lotus team driver. Lotus didn't feel it was necessary to inform us about this and when I asked the people at Lotus about the larger roll bar, they said that they didn't have anymore. At any rate, we did our own thing and qualified on the outside of the front row at a speed of 158.898 miles per hour."—*Dan Gurney*

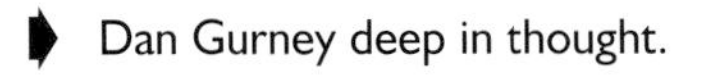

Dan Gurney deep in thought.

After wrecking his rear engine car, Jim Hutubise wound up qualifying the Kurtis Novi that he had driven so spectacularly in 1963. Hurtubise completed only one lap of the race before retiring with transmission problems.

Jim Clark became the first driver to qualify at over 160 miles per hour. Clark had a four lap average of 160.729 miles per hour with a best lap of 160.973 miles per hour. This Lotus 38 was designed by Len Terry and featured a fully enclosed monocoque chassis. Clark led all but 10 laps (Foyt led those) of the race. Clark won the race with a record average speed of 150.686 miles per hour and finished almost two minutes ahead of second place Parnelli Jones.

Canadian driver Billy Foster gets a rather large fan telegram after becoming the first Canadian driver to qualify for the 500.

Al Unser tries out the Lola that Parnelli Jones had as a backup car. Unser wound up qualifying the Lola that A.J. Foyt had as a backup and finished ninth in the race.

A. J. Foyt became the second driver to qualify at over 160 miles per hour when he broke all existing track records by qualifying at a four lap average of 161.233 miles per hour. Foyt had a fast lap of 161.985 miles per hour and won the pole position. On race day, Foyt led the race on several different occasions, but went out of the race with rear end problems. This Lotus 34 had been a spare in 1964 and was not run at Indianapolis.

Bobby Johns finished in seventh place after starting 22nd.

Dan Gurney's Lotus 34 Ford carried sponsorship from the giant Japanese motorcycle company, Yamaha. Gurney used some Yamaha fuel injection improvements in his Ford-Hilborn injection system and they helped to enhance his engine performance.

Mario Andretti at speed.

Lloyd Ruby was typical of the long time roadster drivers who was able to quickly adapt to the new rear engine cars. Ruby drove this Halibrand-Ford to 11th place in the race even though he didn't finish due to a blown engine late in the race. "The reason that so many of our drivers were able to acclimate themselves so quickly to the rear engine car was because, once you are in the automobile, you tend to forget whether the engine is in front of you or behind you. Other than the sitting position, which develops in time, you work on getting the car to handle the way you want it to. Once that happens, you are very comfortable in the car. The one big difference is that you are sitting further up in the chassis of a rear engine car and it's harder to get as good a feel for the car sitting in the middle, as you do when you were sitting in the back of the front engine car. Good racing drivers are capable of making that transition without too much problem."—*Rodger Ward*

Bobby Johns at speed in the short chute between Turn 1 and 2. These short chutes were the best places to shoot from, but you had to have the right equipment to get good shots.

Parnelli Jones really took to his Lotus 34 once he got used to it. "After the 1964 Indianapolis race, I won Trenton and Milwaukee with the Lotus and, at the end of the year, I was given the Lotus that Jimmy put on the pole and Foyt was given one of the spare cars. Once I got used to that car I really loved it and I ran second to Jimmy at Indianapolis in that car."—*Parnelli Jones*

Dan Gurney wasn't very happy with the performance of his Lotus 38 on race day. "I wasn't real happy with the way the car ran during the race. In trying to work around our anti roll bar problem, we made some mistakes in preparation that a more experienced team wouldn't have made, but at the time, we were neophytes in this type of racing. During the race the car was pushing so badly that, coming out of Turn 3 with my foot on the floor, I could turn the wheel all the way to the left to the stop, and the thing just kept going straight toward the wall. It certainly wasn't a very comfortable feeling. Parnelli started behind me and he later asked me 'What in the hell were you doing out there?' I told him 'I just wanted you to know that I was having a little problem and you weren't going to pass me.' He said 'You were a damn fool.'—*Dan Gurney*

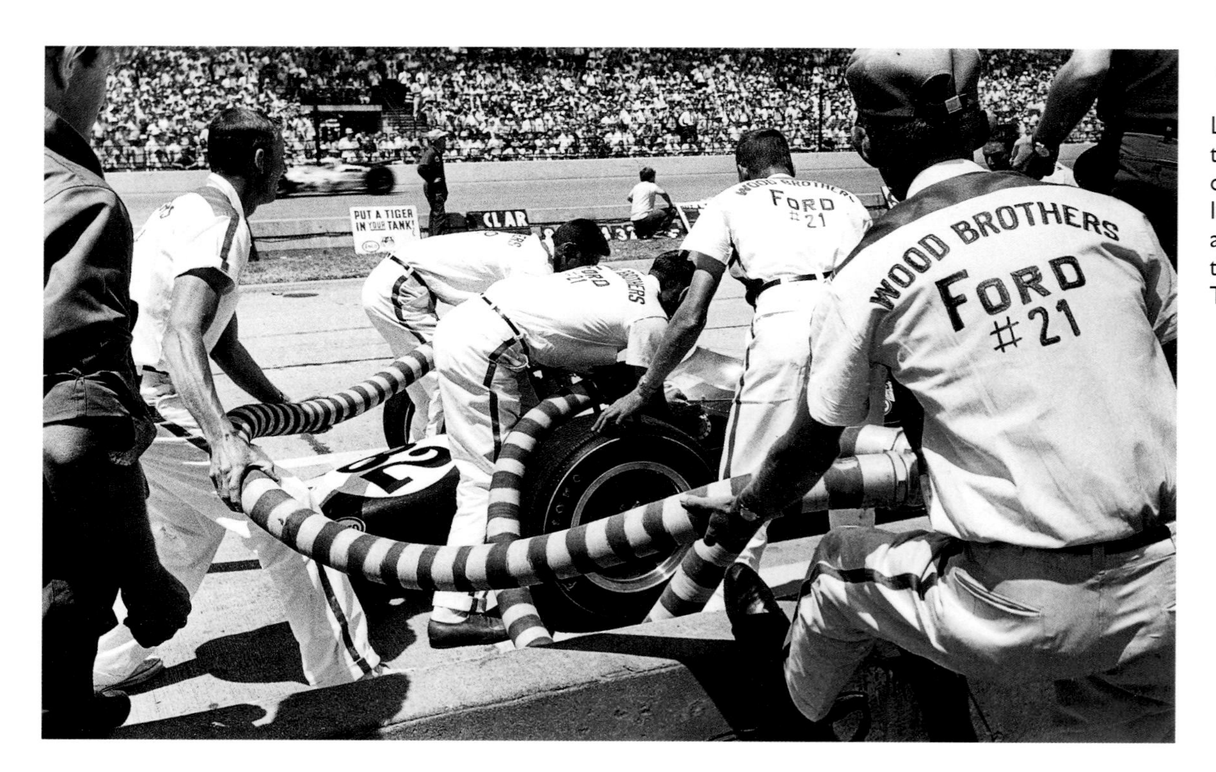

In 1965, the famed Wood Brothers pit crew was brought in to handle the pit stops for Lotus. The Wood Brothers crew were known throughout NASCAR as the fastest, most efficient crew in the business and they proved it at Indianapolis. Note the fuel hose painted to look like a tiger tail. The Lotus effort was using Enco fuel and their advertising slogan was "Put A Tiger In Your Tank." See the pit board in the background.

Dan Gurney was running fourth when his timing gears broke on the 43rd lap. "We finally went out of the race with broken timing gears, and there went opportunity number three to win the race. I guess it could all be called an educational process. Reflecting back, there were many times that, while in the race I said to myself 'If I just had a few moments to stop in the pits and make a few adjustments, I could win this damn thing.' But I guess everyone says that. It's funny, you think that with the full month of May, you have so much time to prepare the car, but by the end of May, you seemed to have run out of time. We really didn't do any testing in those days and that's why the roadsters made such small incremental changes over the years."—*Dan Gurney*

Rookie driver Gordon Johncock finished fifth in this Watson built roadster. It would be the last time that a front engine car would ever finish in the top ten at Indianapolis.

Jim Clark and crew pose for a victory photo the day after the race.

A thrilled Jim Clark and Colin Chapman take a victory lap in the pace car. This was the first victory for a rear engine car at Indianapolis and it began a trend that continues to this day.

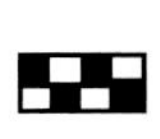

Mario Andretti was the fastest man at Indianapolis all month so it should have been no surprise when he took the pole position with a new record speed of 165. 899 miles per hour. Andretti went out of the race on the 27th lap with a blown engine. The car was officially listed as a Dean Van Lines Hawk with a Brabham by Brawner chassis. A real mouthful.

1966 THE EAGLE ARRIVES AND THE ROADSTER DEPARTS

IN 1966, Dan Gurney arrived at the speedway with his beautiful Len Terry designed Eagle-Ford. Of the five Eagle's entered in the race, all five qualified. Not bad for a first effort, and it was a preview of things to come. Gurney's effort was only outdone by the Gerhardt chassis which qualified six entries.

Rear engine cars comprised 32 spots on the starting grid, while only one roadster managed to make the show. How things changed in just four years.

By 1966, the Ford engine had become dominant with 23 Ford-powered cars in the starting field against just ten Offys.

Goodyear tires had taken a big bite out of the Firestone domination by qualifying 16 of their cars while Firestone qualified 17.

Another stellar crop of rookies showed up at the speedway in 1966. They were led by Graham Hill, who won the race, and Jackie Stewart who almost won the race. Other rookies included Mel Kenyon, Gary Gongdon, Cale Yarborough, Carl Williams, and Larry Dickson.

Most of all, the 1966 race will be remembered for the bitterly disputed race finish between the crews of Andy Granatelli, John Mecom and the USAC officials. Some scoring errors caused both Jim Clark and Graham Hill to arrive at Victory Lane, thinking that they had won the race. As Hill stated so well afterward "I drank the milk, mate." A review of the films and tapes proved that he was entitled to do just that.

▲ 1966 will be remembered for the Indianapolis debut of the Len Terry designed AAR Eagles and five of these beautiful cars qualified for the race in their first year of production. The Eagles would make a considerable impact on USAC racing for years to come. "In 1966 we went to Indianapolis with our first Eagle. Len Terry, who had designed the Lotus team cars for Indianapolis, designed our car. The Eagle was an excellent car for the time. That car had more fuel capacity than the other cars, the chassis was superior to what was being run at the time, the aerodynamics were real close, and it was one of the best looking cars ever built."—*Dan Gurney*

Three of Europe's best drivers made a big impression in the 1966 race. As luck would have it, these three men would be running in the first three positions at the end of 190 laps. But as we all know, the race is never over until the checkered flag falls, even with ten laps to go.

Graham Hill

Jackie Stewart

Jim Clark

Instead of getting smaller and sleeker, the Novi that appeared at the speedway in 1966 was bigger and bulkier. Greg Weld was the assigned driver and this would be the fabled Novi's last appearance at Indianapolis.

One of the more interesting creations to appear in 1966 was a twin engine (one mounted in the front and one mounted in the rear) Porsche powered Huffaker driven by Bill Cheesbourg. Unfortunately, the car couldn't get up to qualifying speed.

Dan Gurney and former motorcycle champion Joe Leonard converse while waiting to practice. Gurney gave the future USAC Champion his first ride at Indianapolis in 1965 and Leonard continued to drive for All American Racers in 1966.

Top view of the AAR Eagle driven by Roger McClusky.

Top view of the last real roadster to qualify for the Indianapolis 500. Bobby Grim drove the turbocharged Offy powered Watson roadster.

Top view of the Lotus 38 driven by Jim Clark.

Graham Hill and Eric Broadley, builder and designer of the Lola Cars that Hill and Jackie Stewart were driving, have a brief discussion before practice.

Dan Gurney accommodates his fan club, something you don't see much of anymore in big time racing.

Jackie Stewart prepares to practice, as chief mechanic George Bignotti (left) directs changes being made by the crew.

As Dan Gurney prepares to qualify, the tension shows on everyone's face.

Jackie Stewart (left), Graham Hill, and Eric Broadley (right) in conference before qualifying.

Greg Weld's attempt to qualify the Novi wound up in a chance meeting with the wall. That crash ended the Novi legacy at Indianapolis.

Jim Clark sees the humor in Colin Chapman's comment.

Jim Clark prepares to qualify his Lotus 38 Ford. Clark qualified for the middle of the front row with an average speed of 164.144 miles per hour. Only Mario Andretti went faster.

A.J. Foyt wrecked his Coyote-Ford just before qualifications and had to fall back upon his tried and true Lotus 34 Ford.

Jackie Stewart (left) and car owner John Mecom of Houston, Texas switch hats.

Legendary Dean Van Lines chief mechanic Clint Brawner accepts the plaudits for a job well done after Mario Andretti's record setting pole position speed.

Bobby Grim qualified the Racing Associates roadster in 31st position with a speed of 158.367 miles per hour.

Unfortunately, Grim was involved in the first lap crash that eliminated 11 cars before the race was one lap old. This was truly the roadster's swan song.

Dan Gurney (31) leads Jerry Grant (88) during Carburetion Day practice runs. Both drivers are driving the new AAR Eagle-Ford. Gurney's clutch problems kept him from qualifying up front and he qualified 19th. This led to serious problems. "We never got a chance to prove ourselves in the race because we got eliminated in that huge crash at the start. If we hadn't had clutch problems during qualifying, we wouldn't have been in a position to be eliminated in that wreck."—*Dan Gurney*

A.J. Foyt's Lotus 34 Ford was also eliminated in the first lap crash. Foyt blamed Graham Hill for starting the accident when Hill supposedly backed off the throttle after taking the green flag.

George Snider qualified Foyt's Lotus-Ford in the third qualifying position. Snider was running third when he was involved in a crash on the 22nd lap.

Jim Clark was never out of contention for his second Indianapolis win, but it wasn't to be. Clark had a close call when he spun his Lotus while holding a 22-second lead, but good fortune was smiling because Clark did not hit anything and he kept going to finish in a disputed second spot.

Jackie Stewart almost won the Indianapolis 500 on his first try but engine failure intervened.

Graham Hill took the lead on the 192nd lap after his Mecom Lola team mate Stewart pushed his car into the pits with a blown engine.

Jackie Stewart won a lot of plaudits for the way he handled his very disappointing, late lap departure from the race. Former well known ABC Wide World Of Sports announcer Chris Shankel conducts the Stewart interview.

Graham Hill won the Indianapolis 500 on his first try but there was a scoring dispute over who really won, between Team Lotus and USAC, that is still argued today.

Jim Clark, thinking that he had won, pulled up to victory lane only to find it occupied by Graham Hill. Team Lotus scorers had Clark winning the race while USAC scorers showed that Hill won. Andy Granatelli can be seen arguing with a USAC official in the background.

After the race, Colin Chapman and Andy Granatelli discuss the reasoning behind the official protest that they have filed with Chief Steward Harlan Fenglar. After a review of films and tapes, Hill's win was allowed to become official.

A.J. Foyt celebrates his third win with his wife Lucy.

1967
SILENT SAM SHOCKS THE ESTABLISHMENT

IN 1967, the turbine car made a serious impression on everyone who saw it run. It created a controversy that would carry on for two years before the foolish, old men of USAC finally effectively banned it. Turbine cars had appeared at the speedway before, but none of them had ever proved to be competitive until Granatelli showed up with his creation.

For the first time in history, there were no front engine cars in the starting field. There were 32 rear engine cars and one side engine car (the turbine). The Eagle was the winner in the chassis department with six entries and Ford had the most engines in the race with 25. The turbo Offy began to make an impression with seven entries and there was the lone turbine engine.

Grand Prix driver Denis Hulme was "Rookie Of The Year" and other rookies included Jochen Rindt, Wally Dallenbach, and LeeRoy Yarbrough.

The 1967 race was one to be remembered. Parnelli Jones provided most of the memories even though A.J. Foyt won his third 500 in six years.

◀ In 1967, the loudest noise heard at Indianapolis was by a car that made no noise at all. The old saying 'That silence is deafening' certainly applied to Parnelli Jones' performance during the month of May. Note that the turbine engine is mounted next to the driver and that a flap appears on the back of the car were the number 40 is. That flap is an air brake.

▼ Richie Ginther (left) and Dan Gurney find time to talk before practice starts. Ginther never felt comfortable on the big oval tack and retired from racing before qualifying began. Jerry Grant took over the car and put it in the race on the final day of qualifying. Grant finished the race in 20th position after suffering a burned piston.

Top NASCAR driver, Cale Yarborough was one of several stock car drivers to drive in the Indianapolis 500 during this period. LeeRoy Yarbrough, Donnie Allison and Bobby Allison were the others. Yarborough started his Vollstedt-Ford in 20th position and was running 11th when a blown tire caused him to spin on the 176th lap. He finished in 17th position.

Parnelli Jones goes out for practice under the watchful eye of Andy Granatelli. The STP Day-Glo Orange was hard to miss wherever you were.

Mario Andretti and the task at hand.

Graham Hill and Chief Mechanic Jim Endruweit discuss the handling of the unique looking Lotus 42F. The car was found to handle much better than the 38 because of its longer wheelbase.

Jim Hurtubise built and drove this Mallard roadster in 1967 and 1968. In 1967, Hutubise was bumped from the starting field but in 1968 he managed to qualify the last roadster type car to ever run in the race. Starting in 30th position, Hurtubise went out of the race on lap 9 with a burned piston. The car was powered by a turbocharged Offy. "Hurtubise was one of the few guys who had a problem with the change over and I don't think that it was because of a lack of ability, I think it was a lack of brains and stubbornness. He just couldn't come to grips with the fact that the day of the roadster was over."—*Rodger Ward*

Jochen Rindt, Denis Hulme, Richie Ginther, and Ronnie Bucknum hold a special driver's meeting.

Jim Clark's well-used Lotus 38 rolls out for practice.

Is Jim Hurtubise (right) suggesting that Jim Clark might try sprint car racing? Hurbubise had about the same reputation on dirt that Clark did on pavement.

Dan Gurney brought rising Grand Prix star Jochen Rindt to Indianapolis to drive one of his Eagles. Rindt would return to the speedway in 1968 and 1969.

Bob Bondurant (left) and Chris Amon converse in the pit lane. Bondurant did not run and Amon, not feeling comfortable in this type of racing environment, returned to Europe before qualifying.

Parnelli Jones passes one of the Ford powered cars at the start-finish line.

Eric Broadley and Jackie Stewart enjoy a seat on the wall during a break in the action.

Looking more like a meeting of the Grand Prix Drivers Association than the Indianapolis 500, Jackie Stewart (far left), Ronnie Bucknum, Chris Amon, Colin Chapman, Denis Hulme, and Jim Clark hold an informal chat in pit lane.

Turbocharger fire during practice. The driver can be seen making a fast exit through the flames.

Parnelli Jones meets the fans and signs autographs.

Graham Hill and crew pose for the photographers after qualifying the Lotus 42F on the final day of qualifying. Andy Granatelli and Colin Chapman are in the red jackets behind Hill.

This Mongoose-Ford built by Dave Laycock was qualified by George Snider. The car started tenth and was running 24th when rain stopped the race. Lloyd Ruby took over the car on the second day and was involved in a crash on the 100th lap. Ruby was unhurt and the car was classified as finishing in 26th place.

By 1967 the Lotus 38s were well used up and Clark could only qualify fast enough for 16th starting position. On race day, Clark was running 11th when the race was halted after 18 laps due to rain. When the race was restarted the next day, Clark suffered engine failure on the 35th lap and was classified 31st overall.

Colin Chapman (left), Andy Granatelli, Jim Clark, and Graham Hill (in car) have a few words before practice.

This odd shaped Lotus 42F was supposed to accommodate a BRM H16 engine for Indianapolis, but that engine never materialized for the race and a Ford engine was installed. Graham Hill qualified on the second weekend and started in 31st position. Hill went out of the race after 23 laps with engine failure and was placed in 32nd position while Clark was placed 31st. The whole 1967 event turned out to be a major embarrassment for Team Lotus.

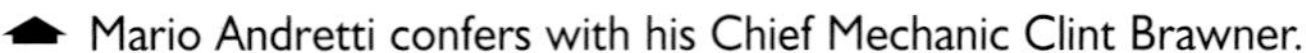

Mario Andretti confers with his Chief Mechanic Clint Brawner.

The two "Great Scots" enjoy a relaxed moment. Between them Jackie Stewart (left) and Jim Clark earned five world championships during their racing careers.

Parnelli Jones (left) and Andy Granatelli confer about the unique STP turbine car. Starting in sixth position, Parnelli Jones blew by Mario Andretti on the back straight to take the lead on the first lap. Because of the turbine car's amazing race day performance, many claimed that Parnelli had been "sand bagging" all month. "Everyone always accuses me of sand bagging at Indianapolis in 1967. The truth is, I did not hold back. I was thinking about ending my career in Indy car racing at that time so I would have had no reason to hold back. Besides USAC couldn't have changed the rules if I could have run 200 miles per hour, so why would I have held back? What happened with the turbine car was that it had tremendous torque. It would really jump across the short chutes and it would jump out on the straight, but it would fall flat on it's ass at the end of the straight away and everyone that I had passed earlier would just drive by me. You have to understand that when the other guys were out practicing, they were running with light loads of fuel and a little bit of nitro. I knew that when they took the nitro out and put 75 gallons of fuel in their cars on race day that they might not be able to drive back by me as they had done in practice and, basically, that's what happened. In the seven years that I ran at Indianapolis, my starting position in 1967 was the worst one that I ever had. When I passed Mario for the lead on the first lap, I glanced over at him and he gave me the finger."—*Parnelli Jones*

Jochen Rindt goes out to practice. Rindt used a stock block Ford engine with Gurney-Weslake heads and went out of the race when he broke a valve on the 108th lap.

A.J. Foyt was one of the greatest American drivers ever and you could never count him out of contention in any type of racing equipment or on any type of racing circuit.

Dan Gurney and his crew stand by his Eagle-Ford prior to the start of the race. "The 1967 Eagle was the best and most beautiful car that I ever drove at Indianapolis."—*Dan Gurney*

▲ Dan Gurney started in the middle of the first row and ran in the front for most of the race. He went out of the race on the 160th lap with a burned piston and finished 21st. "In 1967 we qualified in the middle of the front row next to Mario Andretti and we even held the track record (167.942 miles per hour) for about three minutes until Mario smoked us. When the race started, Parnelli blew by all of us on the first lap and was long gone. Was he sand bagging that year? You'd have to ask him, but I think there's no question that he was. When the rain came and they stopped the race, we managed to make a small adjustment that allowed us to run away from everyone in the field except for Parnelli. We were the fastest of the rest behind Parnelli's turbine. No one could have caught Parnelli that day and the only way that he was going to lose that race, was the way that he did, through mechanical failure. We probably could have won that race, as it turned out, but we dropped a valve. That was our own fault, but that's the way things sometimes are."—*Dan Gurney*

Mario Andretti had been running very strong all month and he set new one (169.779 miles per hour) and four (168.982 miles per hour) lap records during qualifying. During the race, Mario ran into clutch trouble early on, and finally he had the right front wheel come off on the 59th lap. This classified him in 30th position.

One of A.J. Foyt's well coordinated pit stops.

▲ Gordon Johncock (3) battles Lloyd Ruby (25) for position early in the race. Johncock finished 12th while Ruby bent some of his valves at the start of the race and went out after three laps.

After leading 171 laps, Parnelli Jones went out of the race with three laps to go. A bearing failed in the gear box causing the turbine powered car to lose power. Andy Granatelli (left foreground) and Vince Granatelli (next to Parnelli) appear to be in a state of shock.

A.J. Foyt heads for victory lane for the third time in seven years.

Mario Andretti started a lot of rumors when he took a test ride in one of the Lotus-Turbines. Andretti got up over 167 miles per hour, but he didn't switch teams and stayed with the Clint Brawner Hawk.

1968
EAGLE WINS AND TURBINE FAILS AGAIN

THE turbines were back in 1968 and not even the rule changes implemented on the turbine air inlet area by USAC could stop those cars from being the fastest on the track during the month of May. In their undying wisdom, USAC dictated that the turbine's air inlet area on the Pratt & Whitney engines be restricted to 15.99 square inches as opposed to 21.90 square inches that the turbines were allowed to run in 1967. Granatelli claimed that this reduced the horsepower from 550 in 1967 to 450 in 1968. But it didn't matter because bad luck would strike again.

Just when everyone thought there would never be another front engine car on the starting field, Jim Hutubise jumped up and qualified his front engine, turbocharged, Offenhauser roadster in 30th starting position. This would be the last time a front engine car would ever qualify for the Indianapolis 500.

The crop of 1968 rookie drivers included Ronnie Bucknum, Jim Malloy, Mike Mosely, Sammy Sessions, Gary Bettenhausen, and "Rookie Of The Year" Bill Vukovich.

Bobby Unser would become the first driver to win the Indianapolis 500 in an AAR Eagle chassis. Those chassis would also finish second and fourth. The turbine would fail again with victory within its grasp, and after the race, it would be effectively banned forever.

In April 1968, Graham Hill previewed the four wheel drive Lotus 56 to the press. The car was powered by a Pratt & Whitney turbine engine and weighed in at 1,350 pounds which was almost 400 pounds less than "Silent Sam" weighed in 1967.

A very rare picture of Jim Clark (second from left) with the Lotus 56 at Indianapolis during a March testing session. Sadly, Clark would be killed shortly after this picture was taken, and Parnelli Jones (next to Clark) would have second thoughts about driving the turbine car in 1968. Others in the photo are (from left) Andy Granatelli; William R. McCrary, Director of Racing for Firestone; and Colin Chapman.

As the 1967 STP Turbocar that Parnelli Jones was supposed drive sits parked in the foreground, Mike Spence prepares to take Greg Weld's Lotus 56 (background) out for a quick test. Unfortunately, that test resulted in Spence's death. "The reason that I didn't drive the car in 1968 was that Andy wanted me to drive the old car when Chapman had built the new wedge shaped cars. I didn't think that the old car was safe so I didn't drive it. I was proven right when Joe Leonard destroyed the car by spinning it and putting it into the wall. As you know, Joe got into one of the newer wedge cars, qualified on the pole, and damn near won the race If I had never won Indianapolis, you would have never kept me out of the car."—*Parnelli Jones*

◀ Mike Spence was killed when the tire, seen in the cockpit area, hit him in the head after he hit the wall. After this tragedy, Colin Chapman left the speedway saying that, after losing both Clark and Spence exactly one month apart, his heart was not in running this race. A USAC committee inspected the wreck and found no mechanical failure, although several days later, they found that certain Lotus suspension and steering parts were made of the wrong material. Granatelli immediately had new parts made to comply with USAC standards.

▼ Jim Hurubise qualified his front engine, turbocharged Offy Mallard on the last day of qualifying. Hutubise started in 30th position and went out of the race on the ninth lap with a burned piston.

As Johnny Rutherford races down the front straight, his crew checks his lap times.

Jocken Rindt qualified this Repco-Brabham V-8 powered Brabham on the second day of qualifying. Rindt started 16th and went out of the race after five laps with a broken piston.

A very rare picture of Bruce McLaren in his only appearance as a driver at Indianapolis. McLaren and Denny Hulme were to be the drivers of the Shelby turbine cars built by Ken Wallis. Wallis, who had built the turbine car that Parnelli Jones had driven in 1967, failed in this attempt and the cars were withdrawn after failing to reach competitive speeds. Hulme went on to drive one of Gurney's Eagles and finish fourth while McLaren went home.

Joe Leonard (60), Bobby Unser (3), Mario Andretti (2), Lloyd Ruby (25), Roger McCluskey (8), Graham Hill (70), Al Unser (24), and A.J. Foyt (1) lead the rest of the field to the first turn.

The stripped aluminum, monocoque chassis of one of the Lotus 56 Turbocars sit in the garage undergoing a rebuild.

Jack Brabham (right) and long time colleague Ron Tauranac discuss the Brabham Indianapolis car that Jochen Rindt would drive. Brabham was not entered as a driver in 1968, but did some of the testing of his car during the month of May.

Johnny Rutherford was one of racing's really good guys. Rutherford would finish 18th in his AAR Eagle-Ford in 1968, but his day would come and he would win three Indianapolis 500 races before he retired from racing.

1968 was not A.J. Foyt's year. Driving one of his Coyote-Fords, Foyt qualified eighth and was running in sixth position when the rear end gears failed on the 87th lap.

Art Pollard is besieged by photographers, crew, and officials as he prepares to qualify. Jack Brabham in the driver's uniform on the right is also an interested spectator. Pollard qualified at a speed of 166.297 miles per hour and started in 14th position.

▲ Joe Leonard is interviewed after winning the pole position. Surrounding Leonard are Joe Granatelli, Parnelli Jones, and Andy Granatelli who are listening to what Leonard has to say. Leonard became the first driver to ever qualify at over 170 miles per hour with a record qualifying speed of 171.559 miles per hour and a record one lap speed of 171.953 miles per hour.

▶ Parnelli Jones, Andy Granatelli, and Joe Leonard celebrate after Leonard's record qualifying run. Parnelli was responsible for tuning the chassis of Leonard's car. He obviously knew what he was doing.

Billy Vukovich (98) and Graham Hill (70) race wheel to wheel through the first turn.

Bobby Grim (6) leads Mario Andretti (2) through the short chute. Grim finished 10th and Andretti lasted only two laps before engine trouble set in.

Dan Gurney (48) leads Sammy Sessions (94) and Bobby Grim (6). "In 1968 we went to Indianapolis with our Ford stock block engine with the Gurney Weslake cylinder heads, instead of the Ford four-cammer engine. Our 1968 car was pretty good but it wasn't as good a car as the 1967 Eagle. Tony Southgate designed that car and it wasn't as reliable as the earlier one. We qualified for the 1968 race on 40% nitro and ran the race on 15% nitro. That car was a good car in the race, fun to drive, and we finished second that year to Bobby Unser who won the race in one of our Eagles owned by Bob Wilke's Leader Card team. That was the Eagle's first win at Indianapolis so we were quite happy with our sweep of the first two places and that race was probably my favorite Indianapolis race. We won a lot of races with that 68 Eagle."—*Dan Gurney*

Al Unser's Lola hit the wall on the 41st lap and he took the long walk back to the pits.

Andy Granatelli wipes Joe Leonard's visor as he takes on 74 gallons of fuel.

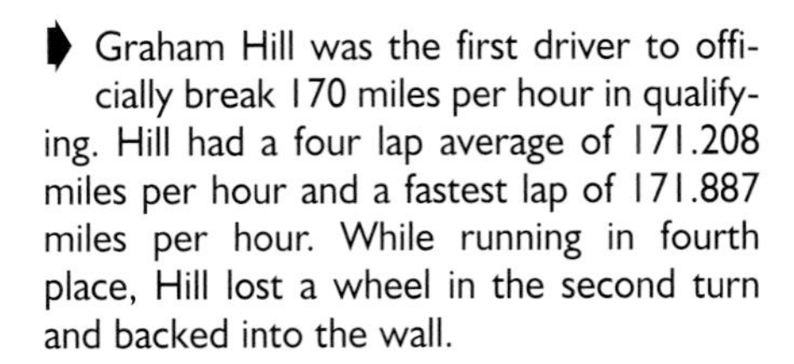

Graham Hill was the first driver to officially break 170 miles per hour in qualifying. Hill had a four lap average of 171.208 miles per hour and a fastest lap of 171.887 miles per hour. While running in fourth place, Hill lost a wheel in the second turn and backed into the wall.

Denis Hulme was able to get a ride with Dan Gurney's team after the Shelby Turbine effort withdrew from the race. Hulme started 20th and was able to finish a fine fourth overall. Hulme's car ran a Ford four cam engine instead of the stock block Ford engine that Gurney ran.

Jim McElreath's Coyote gives up the battle.

Bobby Unser celebrates his first of the three wins that he would have at Indianapolis during his racing career.

Bobby Unser led a total of 127 laps of the 200 lap race. He took the lead for good on the 192nd lap and went on to gain the first win for the Dan Gurney-built Eagle. The Eagles finished first (Unser), second (Gurney), and fourth (Hulme).

Bobby Unser gets one of his many awards for winning the Indianapolis 500.

Brabham teammates Peter Revson (left) and Jack Brabham celebrate their qualifying for the race.

1969
PENSKE ARRIVES AND ANDRETTI PERSEVERES

THE race in 1969 would see the arrival of a new team that, in later years, would dominate the Indianapolis 500 in a way that no other team ever has or ever will. That team was Penske Racing, headed by a youthful Roger Penske and his driver Mark Donohue. The team adapted quickly to this new form of racing and Donohue finished seventh in spite of a lengthy pit stop for a magneto replacement.

The 1969 race would also have one of those very special moments that, once in a while, happen to people who deserve them. Mario Andretti had been among the fastest drivers on the track with his new Lotus 64-Ford. On May 21, Mario was involved in a spectacular crash that left his Lotus broken in two and left Andretti with relatively minor facial burns. The next day Andretti was back on the track in his backup car, and on the first day of qualifying Mario put that car in the middle of the first row next to A.J. Foyt who sat on the pole.

On race day, Andretti's STP Hawk-Ford took the lead on lap 100 and never looked back. It was certainly one of the most popular wins in recent speedway history. Andy Granatelli, Andretti's sponsor, had been robbed of victory the past two years with less then ten laps to go. This year however, Granatelli would get his long sought after victory with a car that USAC couldn't complain about.

It was sad that during that period, USAC was controlled by a bunch of silly, old men who wanted nothing to progress any further than the front engine roadster. These officials went out of their way to ban everything that could make the race more interesting, and it was surprising that they couldn't come up with a reason to ban the rear engine car.

Jim Hurtubise was back for the third year with his front engine, turbocharged Offy-Mallard, but luck wasn't with him. Hutubise was running laps at 163+ during qualifying, but shut off for some unexplained reason and didn't qualify. Had he continued, he would have made the race.

Is it a Lotus or a very good copy? You would be right if you said it was a Gerhardt-Offy driven by Carl Williams and that it finished 25th after starting in 30th position.

When Clint Brawner (right) speaks, Mario Andretti and Colin Chapman listen.

Looking more like a land speed record car then a car built to run at Indianapolis, the turbine powered, Glenn Bryant designed, Al Miller driven creation failed to get up over 156.440 miles per hour. This was not good enough to qualify for the starting field.

Mario Andretti prepares to take a few quick laps in the new Lotus 64 Ford. This car was considered the most complex Lotus ever built. Since USAC had effectively legislated the competitive turbine engine out of existence, a turbocharged Ford V-8 was used along with a modified four wheel drive system. USAC had also, initially, banned the four wheel drive chassis, but they later allowed its use as long as the wheel width was limited to 10 inches. Two wheel drive cars were permitted to use 14 inch rims.

Penske Racing and Mark Donohue made their initial debut at Indianapolis in 1969. The car that they used for that race was a four-wheel drive Lola powered by a turbocharged Offy. The original Penske effort was ridiculed by the Indianapolis regulars, but as we know, the joke would be on them. Penske Racing would become the most dominant team in the history of the speedway.

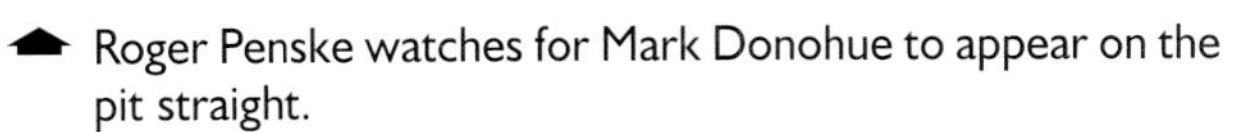

Roger Penske watches for Mark Donohue to appear on the pit straight.

Graham Hill, Colin Chapman, and Mario Andretti look for more speed.

Dan Gurney is ready to make a test run. Gurney qualified in tenth position and started on the inside of the fourth row. "In 1969 we finished second, again, to Mario Andretti who finally got his well-deserved win there. As I remember, we were both limping along at the finish."—*Dan Gurney*

The very sanitary Lotus 64 chassis undergoes a complete check out before being re-assembled.

Mario Andretti was among the fastest drivers during practice with lap speed in excess of 171 miles per hour. Mario felt that he had a good chance to sit on the pole, but it was not to be.

On May 21, Andretti had a huge crash when the right rear hub broke. He hit the wall and slid for 600 feet before catching fire and stopping. Mario had facial burns but walked away from the accident. A USAC investigation showed that poor heat treatment was the cause of the failure, and since there wasn't time to have new hubs made and tested, the Lotus 64s were withdrawn from the race. A battle of ownership of the cars broke out between Granatelli and Chapman and the cars were eventually shipped back to England under wraps. This ended the Lotus effort at Indianapolis.

A. J. Foyt put his Coyote-Ford on the pole with a four lap average of 170.568 miles per hour and a fast lap of 171.625 miles per hour. Foyt did not break Joe Leonard's qualifying record of 171.599 miles per hour set in 1968. Foyt had been unofficially over 172 miles per hour in practice, but track conditions prevented him from reaching that mark in qualifying.

The ever popular Jack Brabham prepares to qualify as Ron Tauranac gives Jack a last bit of encouragement.

Denis Hulme demonstrates his own special technique of checking suspension.

Bobby Unser's new four-wheel drive Lola-Offy started in third position and finished in third position.

It's hard to tell the difference between the first day of qualifying and race day when it comes to the crowd and the confusion around the entrance to Gasoline Alley.

Jack Brabham qualified his Repco-Brabham in 29th position, but lasted only 58 laps because of ignition problems.

Peter Revson is interviewed after he qualified the second team Repco-Brabham in 33rd and last position. During the race, however, Peter moved from 33rd starting position to finish in fifth place. This feat earned him the "Co-Rookie Of The Year" award.

Good friends and former team mates Denis Hulme (left) and Jack Brabham share a good laugh after qualifying.

Chief Steward Harlan Fengler (red hat) talks to Clint Brawner as Mario Andretti prepares to qualify his Brawner-Hawk III-Ford. Jim McGee appears behind Andretti. When Andretti destroyed the Lotus, he jumped into his backup car and started from the middle of the second row.

Roger Penske (yellow shirt) stands by as Mark Donohue prepares for a qualifying attempt. Chief Mechanic Karl Kainhofer stands at the rear of the car. Donohue started fourth and finished seventh. He shared the "Rookie Of The Year" award with Peter Revson.

Mark Donohue is interviewed by Chris Economaki after his qualifying run.

Jim McElreath makes a fast exit after his supercharger blew during the race and ruptured the fuel pickup tank. A small fire resulted, but McElreath escaped unhurt.

▲ A.J. Foyt started from the pole but finished eighth after turbocharger problems caused him to make a 24 minute pit stop.

▼ Peter Revson gives a great display of driving ability by finishing fifth after starting last in the field.

Mark Donohue's crew took ten minutes to replace the Lola's magneto. In spite of the trouble, Donohue still finished a fine seventh overall.

Mario Andretti on his way to his only Indianapolis win. During his driving career, Andretti had the worst racing luck of any top driver at Indianapolis. It seems that he was, generally, the fastest in practice and qualifying and yet during the race, he always had problems.

Mario Andretti, still showing the effects of the burns on his face, accepts the winner's laurels. Andy Granatelli is at the left and Joe Granatelli is behind Andy. By 1969 the winds of change had blown through the Indianapolis race and the turbocharged, rear engine car had emerged as the car to beat. It's still that way today.

INDEX